Luigi Amerio • B. Segre (Eds.)

Sistemi dinamici e teoremi ergodici

Lectures given at the
Centro Internazionale Matematico Estivo (C.I.M.E.),
held in Varenna (Como), Italy,
June 2-11, 1960

Springer

C.I.M.E. Foundation
c/o Dipartimento di Matematica "U. Dini"
Viale Morgagni n. 67/a
50134 Firenze
Italy
cime@math.unifi.it

ISBN 978-3-642-10943-0   e-ISBN: 978-3-642-10945-4
DOI:10.1007/978-3-642-10945-4
Springer Heidelberg Dordrecht London New York

Printed on acid-free paper

Springer.com

CENTRO INTERNATIONALE MATEMATICO ESTIVO
(C.I.M.E)

Reprint of the 1ˢᵗ ed.- Varenna, Italy, June 2-11, 1960

SISTEMI DINAMICI E TEOREMI ERGODICI

CENTRO INTERNATIONALE MATEMATICO ESTIVO

(C.I.M.E)

PAUL R. HALMOS

ENTROPY IN ERGODIC THEORY

ROMA - Istituto Matematico dell' Università - 1960

P.R.Halmos

CONTENTS

P.R.Halmos

P.R.Halmos

PREFACE

Shannon's theory of information appeared on the mathematical scene in 1948; in 1958 Kolmogorov applied the new subject to solve some relatively old problems of ergodic theory. Neither the general theory nor its special application is as well known among mathematicians as they both deserve to be; the reason, probably, is faulty communication. Most extant expositions of information theory are designed to make the subject palatable to non-mathematicians, with the result that they are full of words like "source" and "alphabet". Such words are presumed to be an aid to intuition; for the serious student, however, who is anxious to get at the root of the matter, they are more likely to be confusing than helpful. As for the recent rgodic application of the theory, the communication trouble there is that so far the work of Kolmogorov and his school exists in Doklady abstracts only, in Russian only. The purpose of these notes is to present a stop-gap exposition of some of the general theory and some of its applications. While a few of the proofs may appear slightly different from the corresponding ones in the literature, no claim is made for the novelty of the results. As a prerequisite, some familiarity with the ideas of the general theory of measure is assumed; Halmos's *Measure theory* (1950) is an adequate reference.

Chapter I begins with relatively well known facts about conditional expectations; for the benefit of the reader who does not know this technical probabilistic concept, several standard proofs are reproduced. Standard reference: Doob, *Stochastic processes* (1953). A special case of the martingale convergence theorem is proved by what is essentially Lévy's original method (*Théorie de*

P.R.Halmos

l'addition des variables aléatoires (1937)). The reader who knows the martingale theorem can skip the whole chapter, except possibly Section 9, and, in particular., equation (9.1).

Chapter II motivates and defines information. Standard refe- rence: Khinchin, *Mathematical foundations of information theory* (1957). The more recent book of Feinstein, *Foundations of informa- tion theory* (1958), is quite technical, but highly recommended. The chapter ends with a proof McMillan's theorem (mean convergen- ce); the reader who knows that theorem can skip the chapter after looking at it just long enough to absorb the notation. Almost eve- rywhere convergence probably holds. A recent paper by Breiman (Ann. Math. Stat. 28 (1957) 809-811) asserts it, but that paper has an error; at the time these lines were written the correction has not appeared yet. In any case, for the ergodic application not even mean convergence is necessary; all that is needed is the convergen- ce of the integrals, which is easy to prove directly.

Chapter III studies entropy (average amount of information); all the facts here are direct consequences of the definitions, via the machinery built up in the first two chapters.

Chapter IV contains the application to ergodic theory. In ge- neral terms, the idea is that information theory suggests a new invariant (entropy) of measure-preserving transformations. The new invariant is sharp enough to distinguish between some hitherto indistinguishable transformations (e.g., the 2-shift and the 3- shift). The original idea of using this invariant is due to Kol- mogorov (Doklady 119 (1958) 861-864 and 124 (1959) 754-755). An improved version of the definition is given by Sinai (Doklady 124 (1959) 768-771), who also computes the entropy of ergodic

P.R.Halmos

automorphisms of the torus. The new invariant is in some respects
not so sharp as older ones. Thus for instance Rokhlin (Doklady
124 (1959) 980-983) asserts that all translations (in compact a-
belian groups) have the same entropy (namely zero); he also begins
the study of the connection between entropy and spectrum. Much remains to be done along all these lines.

P.R.Halmos

CHAPTER I. CONDITIONAL EXPECTATION

SECTION 1. DEFINITION. We shall work, throughout what follows, with a fixed probability space

$$(X, \mathcal{S}, P).$$

Here X is a non-empty set, $\mathcal{S}$ is a field of subsets of X, and P is a probability measure on $\mathcal{S}$. The word "field" in these notes is an abbreviation for "collection of sets closed under the formation of complements and *countable* unions". A probability measure on a field of subsets of X is a measure P such that

$$P(X) = 1 .$$

Suppose that $\mathcal{C}$ is a subfield of $\mathcal{S}$ and f is an integrable real function on X. If

$$Q(C) = \int_C f \, dP$$

for each C in $\mathcal{C}$, then Q is a signed measure on $\mathcal{C}$, absolutely continuous with respect to P (or, rather, with respect to the restriction of P to $\mathcal{C}$). The Radon-Nikodym theorem implies the existence of an integrable function f^* , measurable $\mathcal{C}$, such that

$$Q(C) = \int_C f^* \, dP$$

for each C in $\mathcal{C}$. The function f^* is uniquely determined (to within a set of measure zero); its dependence on f and $\mathcal{C}$ is indicated by writing

$$f^* = E(f/\mathcal{C}) .$$

The function $E(f/\mathcal{C})$ is called "the conditional expectation of f with respect to $\mathcal{C}$ ". It is worth while to repeat the characteristic properties of conditional expectation; they are that

(1.1) $\qquad E(f/\mathcal{C})$ is measurable $\mathcal{C}$

P.R.Halmos

and

$$(1.2) \qquad \int_C E(f/\mathcal{C}) \, dP = \int_C f \, dP$$

for each C in $\mathcal{C}$.

SECTION 2. EXAMPLES. If $\mathcal{C}$ is the largest subfield of $\mathcal{S}$,
that is $\mathcal{C} = \mathcal{S}$, then f itself satisfies (1.1) and (1.2), so that
$$E(f/\mathcal{S}) = f .$$
This result has a trivial generalization: since f always satisfies
(1.2) ($\int_C f \, dP = \int_C f \, dP$), it follows that if the field $\mathcal{C}$ is such
that f is measurable $\mathcal{C}$, then

$$(2.1) \qquad E(f/\mathcal{C}) = f .$$

To look at the other extreme, let 2 be the smallest subfield of
$\mathcal{S}$, that is the field whose only non-empty member is X . Since
the only functions measurable 2 are constants, and since the only
constant (in the role of $E(f/\mathcal{C})$) that satisfies (1.2) is $\int_C f \, dP$,
it follows that

$$(2.2) \qquad E(f/2) = \int f \, dP .$$

The constant $\int_C f \, dP$ is sometimes called the absolute (as opposed
to conditional) expectation of f , and, in that case, it is denoted by E(f).

Here is an illuminating example. Suppose that X is the unit
square, with the collection of Borel sets in the role of $\mathcal{S}$ and
Lebesgue measure in the role of P . We say that a set in $\mathcal{S}$ is
"vertical" in case its intersection with each vertical line L in
the plane is either empty or else equal to $X \cap L$. The collection
$\mathcal{C}$ of all vertical sets in $\mathcal{S}$ is a subfield of $\mathcal{S}$. A function f

P.R.Halmos

is measurable $\mathcal{C}$ if and only if it does not depend on its second (vertical) argument; it follows easily that if f is integrable, then

$$E(f/\mathcal{C})(x, y) = \int f(x, u)\, du \ .$$

SECTION 3. ALGEBRAIC PROPERTIES. Conditional expectation is a generalized integral and in one form or another it has all the properties of an integral. Thus, for instance,

$$(3.1) \qquad\qquad E(1/\mathcal{C}) = 1 \ ,$$

where this equation, as well as all other asserted equations and inequalities involving conditional expectations, holds almost everywhere. (To prove (3.1), apply (2.1).) If f and g are integrable functions and if a and b are constants, then

$$(3.2) \qquad E(af+bg/\mathcal{C}) = aE(f/\mathcal{C}) + bE(g/\mathcal{C}) \ .$$

(Proof: if C is in $\mathcal{C}$, then the integrals over C of the two sides of (3.2) are equal to each other). If $f \geqq 0$, then

$$(3.3) \qquad\qquad E(f/\mathcal{C}) \geqq 0 \ .$$

(Proof: if $C = \{x : E(f/\mathcal{C})(x) < 0\}$, then C is in $\mathcal{C}$ and $\int_C f\, dP = 0$; this implies that $P(C) = 0$). It is a consequence of (3.3) that

$$(3.4) \qquad |E(f/\mathcal{C})| \leqq E(|f|/\mathcal{C}) \ .$$

(Proof: both $|f| - f \geqq 0$ and $|f| + f \geqq 0$, and therefore, by (3.2) and (3.3), both $E(-f/\mathcal{C}) \leqq E(|f|/\mathcal{C})$ and $E(f/\mathcal{C}) \leqq E(|f|/\mathcal{C})$).

Conditional expectations also have the following multiplicative property: if f is integrable and if g is bounded and measurable

P.R.Halmos

$\mathcal{C}$, then

$$(3.5) \qquad\qquad E(fg/\mathcal{C}) = E(f/\mathcal{C})g \ .$$

Since the right side of (3.5) is measurable $\mathcal{C}$, the thing to prove is that

$$(3.6) \qquad\qquad \int_C E(f/\mathcal{C})g \ dP = \int_C fg \ dP$$

for each C in $\mathcal{C}$. In case g is the characteristic function of a set in $\mathcal{C}$, (3.6) follows from the defining equation (1.2) for conditional expectations. This implies that (3.6) holds whenever g is a finite linear combination of such characteristic functions, and hence, by approximation, that (3.6) holds whenever g is a bounded function measurable $\mathcal{C}$.

SECTION 4. DOMINATED CONVERGENCE. The usual limit theorems for integrals also have their analogues for conditional expectations. Thus if f, g, and f_n are integrable functions, if $|f_n| \leqq g$ and $f_n \to f$ almost everywhere, then

$$(4.1) \qquad\qquad E(f_n/\mathcal{C}) \to E(f/\mathcal{C})$$

almost everywhere and, also, in the mean. For the proof, write

$$g_n = \sup \ (|f_n - f|, \ |f_{n+1} - f|, \ |f_{n+2} - f|, \dots);$$

observe that the sequence $\{g_n\}$ tends monotonely to 0 almost everywhere and that $g_n \leqq 2g$. It follows that the sequence $\{E(g_n/\mathcal{C})\}$ is monotone decreasing and, therefore, has a limit h almost everywhere. Since

$$(4.2) \qquad \int h \ dP \leqq \int E(g_n/\mathcal{C}) \ dP = \int g_n \ dP ,$$

P.R.Halmos

and since $\int g_n\, dP \to 0$, this implies that $E(g_n/\mathfrak{C}) \to 0$ almost everywhere. Since, finally,

$$(4.3) \qquad |E(f_n/\mathfrak{C}) - E(f/\mathfrak{C})| \leqq E(|f_n - f|/\mathfrak{C}) \leqq E(g_n/\mathfrak{C}) \, ,$$

the proof of almost everywhere convergence is complete.

Mean convergence is implied by the inequality

$$(4.4) \qquad \int |E(f_n/\mathfrak{C}) - E(f/\mathfrak{C})|\, dP \leqq \int |f_n - f|\, dP$$

and the Lebesgue dominated convergence theorem.

SECTION 5. CONDITIONAL PROBABILITY. If A is a measurable set (that is A is in $\mathfrak{S}$) and if

$$f = c(A)$$

(where $c(A)$ is the characteristic function of A), we write

$$E(f/\mathfrak{C}) = P(A/\mathfrak{C}) \, .$$

The function $P(A/\mathfrak{C})$ is called "the conditional probability of A with respect to $\mathfrak{C}$". The characteristic properties of conditional probability are that

$$P(A/\mathfrak{C}) \quad \text{is measurable } \mathfrak{C}$$

and

$$\int_C P(A/\mathfrak{C})\, dP = P(A \cap C)$$

for each C in $\mathfrak{C}$. If A is in $\mathfrak{C}$, then

$$(5.1) \qquad P(A/\mathfrak{C}) = c(A) \, ,$$

and, in any case,

$$(5.2) \qquad P(A/2) = P(A).$$

For this reason the constant $P(A)$ is sometimes called the absolu-

P.R.Halmos

te (as opposed to conditional) probability of A.

The converse of the conclusion (5.1) is true and sometimes useful. The assertion is that if $P(A/\mathcal{E})$ is the characteristic function of some set, say B, then

(5.3) $\qquad\qquad\qquad A \text{ is in } \mathcal{E}$

(and therefore B = A). To prove this, note that

$$\int_C c(B)\ dP = P(A \cap C)\ ,$$

and therefore $P(A \cap C) = P(B \cap C)$ for each C in $\mathcal{E}$. Since $P(A/\mathcal{E})$ is measurable $\mathcal{E}$, the set B itself belongs to $\mathcal{E}$. It is therefore permissible to put C = B and to put C = X - B; it follows that $B \subset A$ and $A \subset B$ (almost), so that B = A (almost).

Just as conditional expectation has the properties of an integral, conditional probability has the properties of a probability measure. Thus if A is a measurable set, then

$$0 \leqq P(A/\mathcal{E}) \leqq 1\ ,$$

and if $\{A_n\}$ is a disjoint sequence of measurable sets with union A, then

$$P(A/\mathcal{E}) = \Sigma_n\ P(A_n/\mathcal{E})\ .$$

SECTION 6. JENSEN'S INEQUALITY. A useful analytic property of integration is known as Jensen's inequality, which we now proceed to state and prove in its generalized (conditional) form.

A real-valued function F defined on an interval of the real line is called *convex* if

$$F\ (ps + qt) \leqq pF(s)\ + qF(t)$$

whenever s and t are in the domain of F and p and q are non-negative

P.R.Halmos

numbers with sum 1. It follows immediately, by induction, that if $t_1,\ldots,t_n$ are in the domain of F and $p_1,\ldots,p_n$ are non-negative numbers with sum 1, then

$$(6.1) \qquad F(\Sigma_{i=1}^{n} p_i t_i) \leqq \Sigma_{i=1}^{n} p_i F(t_i) \ .$$

Suppose now that F is a continuous convex function whose domain is a finite subinterval of $[0, \infty)$, suppose that g is a measurable function on X whose range is (almost) included in the domain of F, and suppose that $\mathcal{C}$ is an arbitrary subfield of $\mathcal{S}$. Jensen s inequality asserts that under these conditions

$$(6.2) \qquad F(E(g/\mathcal{C})) \leqq E(F(g)/\mathcal{C})$$

almost everywhere. Since g is the limit of an increasing sequence of simple functions, and since F is continuous, it is sufficient. to prove (6.2) in case

$$g = \Sigma_A c(A) t_A$$

where the summation extends over the atoms of some finite subfield of $\mathcal{S}$. If g has this form, then

$$F(g) = \Sigma_A c(A) F(t_A)$$

and

$$E(g/\mathcal{C}) = \Sigma_A P(A/\mathcal{C}) t_A \ .$$

Since $E(F(g)/\mathcal{C}) = \Sigma_A P(A/\mathcal{C}) F(t_A)$, the inequality (6.2) is in this case a special case of (6.1).

In the extreme case, $\mathcal{C} = \mathcal{S}$, the conditional form of Jensen's inequality reduces to a triviality $(F(g) \leqq F(g))$; in the other extreme case, $\mathcal{C} = 2$, it becomes the classical absolute Jensen's inequality

P.R.Halmos

$$F(\int g \ dP) \leqq \int F(g) \ dP \ .$$

SECTION 7. TRANSFORMATIONS. Later we shall need to know the effect of measure-preserving transformations on conditional expectations and probabilities. Suppose therefore that T is a measure-preserving transformation on X; this means that if A is measurable, then $T^{-1}A$ is measurable and

$$P(T^{-1}A) = P(A) \ .$$

(For present purposes T need not be invertible).

If $\mathcal{C}$ is a subfield of $\mathcal{S}$, then

$$T^{-1}\mathcal{C}$$

is the collection (field) of all sets of the form $T^{-1}C$ with C in $\mathcal{C}$; if f is a function on X, then fT is the composite of f and T. The basic change-of-variables result is that if f is integrable, then

$$\int_C f \ dP = \int_{T^{-1}C} fT \ dP$$

for each measurable set C. If, in particular, C is in $\mathcal{C}$, then

$$\int_{T^{-1}C} E(fT/T^{-1}\mathcal{C}) \ dP = \int_{T^{-1}C} fT \ dP$$

$$= \int_C f \ dP = \int_C E(f/\mathcal{C}) \ dP$$

$$= \int_{T^{-1}C} E(f/\mathcal{C})T \ dP \ .$$

Since both $E(fT/T^{-1}\mathcal{C})$ and $E(f/\mathcal{C})T$ are measurable $T^{-1}\mathcal{C}$, it follows that

$$(7.1) \qquad E(ft/T^{-1}\mathcal{C}) = E(f/\mathcal{C})T \ .$$

Since $c(A)T = c(T^{-1}A)$, this implies that

$$(7.2) \qquad P(T^{-1}A/T^{-1}\mathcal{C}) = P(A/\mathcal{C})T \ .$$

P.R.Halmos

SECTION 8. LATTICE PROPERTIES. The next item of interest is the dependence of $E(f/\mathcal{C})$ on $\mathcal{C}$. The collection of all subfields of $\mathcal{S}$ has a reasonable amount of structure; it is partially ordered (by inclusion), and, in fact, it is a complete lattice. (The infimum of two subfields $\mathcal{B}$ and $\mathcal{C}$ is just their intersection, and their supremum $\mathcal{B} \vee \mathcal{C}$ is the field they generate; similar assertions hold for the infimum and supremum of any family of subfields). It might therefore be hoped that the dependence of $E(f/\mathcal{C})$ on $\mathcal{C}$ exhibits some algebraically pleasant behavior, such as monotony or additivity. Nothing like this is true.

If, for instance, $\mathcal{B}$ and $\mathcal{C}$ are subfields of $\mathcal{S}$ with $\mathcal{B} \subset \mathcal{C}$, then the best that can be said is that

$$(8.1) \qquad E(E(f/\mathcal{C})/\mathcal{B}) = E(f/\mathcal{B})$$

and

$$(8.2) \qquad \int |E(f/\mathcal{B})| \, dP \leqq \int |E(f/\mathcal{C})| \, dP .$$

To prove (8.1) observe that if B is in $\mathcal{B}$, then B is in $\mathcal{C}$, and therefore

$$\int_B E(E(f/\mathcal{C})/\mathcal{B}) \, dP = \int_B E(f/\mathcal{C}) \, dP$$

$$= \int_B f \, dP = \int_B E(f/\mathcal{B}) \, dP .$$

To prove (8.2), write $B = \{x: E(f/\mathcal{B})(x) \geqq 0\}$. Since, clearly, B is in $\mathcal{B}$, it follows that

$$\int_B E(f/\mathcal{B}) \, dP = \int_B E(f/\mathcal{C}) \, dP \leqq \int_B |E(f/\mathcal{C})| \, dP$$

and

$$-\int_{X-B} E(f/\mathcal{B}) \, dP = -\int_{X-B} E(f/\mathcal{C}) \, dP \leqq \int_{X-B} |E(f/\mathcal{C})| \, dP.$$

In case $f = c(A)$, the equation (8.1) becomes

P.R.Halmos

$$(8.3) \qquad\qquad E(P(A/\mathcal{C})/\mathcal{B}) = P(A/\mathcal{B}) .$$

SECTION 9. FINITE FIELDS. Some insight can be gained by studying the finite subfields of $\mathcal{S}$. A finite subfield $\mathcal{B}$ of $\mathcal{S}$ is atomic. This means that $\mathcal{B}$ has a finite number of (necessarily disjoint) atoms whose union is (almost) equal to X; an atom is a set of positive measure in $\mathcal{B}$ that includes no subset of strictly smaller positive measure in $\mathcal{B}$. A function is measurable $\mathcal{B}$ if and only if it is a constant on each atom of $\mathcal{B}$. It follows easily that if f is integrable, then on each atom B of $\mathcal{B}$

$$E(f/\mathcal{B}) = \frac{1}{P(B)} \int_B f \, dP .$$

A convenient way of expressing this fact is to write

$$E(f/\mathcal{B}) = \Sigma_B \frac{c(B)}{P(B)} \int_B f \, dP ,$$

where the summation extends over all the atoms of $\mathcal{B}$. In case f = c(A), this becomes

$$P(A/\mathcal{B}) = \Sigma_B c(B) \frac{P(A \cap B)}{P(B)} .$$

Suppose next that $\mathcal{B}$ and $\mathcal{C}$ are two finite subfields of $\mathcal{S}$. It follows, of course, that $\mathcal{B} \vee \mathcal{C}$ is a finite subfield of $\mathcal{S}$, and the atoms of $\mathcal{B} \vee \mathcal{C}$ are the non-empty sets of the form $B \cap C$, where B is an atom of $\mathcal{B}$ and C is an atom of $\mathcal{C}$. More generally, if $\mathcal{B}$ and $\mathcal{C}$ are subfields of $\mathcal{S}$ such that $\mathcal{B}$ is finite, then every element of $\mathcal{B} \vee \mathcal{C}$ is obtained as follows : for each atom B of $\mathcal{B}$ let C_B be an element of $\mathcal{C}$ and form the union $\bigcup_B (B \cap C_B)$, extended over all atoms of $\mathcal{B}$. Something useful can be said about conditional expectations and probabilities in this case: the assertion

18

P.R.Halmos

is that if f is integrable, then on each atom B of $\mathcal{B}$

$$E(f/\,\mathcal{B}\vee\mathcal{C}\,) = \frac{1}{P(B/\,\mathcal{C}\,)}\ E(c(B)f/\,\mathcal{C}\,).$$

Equivalently

$$(9.1) \qquad E(f/\,\mathcal{B}\vee\mathcal{C}\,) = \Sigma_B\ \frac{c(B)}{P(B/\,\mathcal{C}\,)}\ E(c(B)f/\,\mathcal{C}\,),$$

where the summation extends over all the atoms of $\mathcal{B}$. In case
f = c(A), this becomes

$$(9.2) \qquad P(A/\,\mathcal{B}\vee\mathcal{C}\,) = \Sigma_B\ c(B)\ \frac{P(A\cap B/\,\mathcal{C}\,)}{P(B/\,\mathcal{C}\,)}\ .$$

To prove (9.1), observe that both sides are measurable $\mathcal{B}\vee\mathcal{C}$;
it is therefore sufficient to prove that whenever B is an atom of
$\mathcal{B}$ and C is in $\mathcal{C}$, then the integrals of the two sides of (9.1)
over $B\cap C$ are equal. The proof is the following straightforward
computation:

$$\int_C c(B)\ \frac{E(c(B)f/\,\mathcal{C}\,)}{P(B/\,\mathcal{C}\,)}\ dP = \int_C e(c(B)\ \frac{E(c(B)f/\,\mathcal{C}\,)}{P(B/\,\mathcal{C}\,)}/\,\mathcal{C}\,)\ dP$$

$$= \int_C\ E(c(B)f/\,\mathcal{C}\,)\ dP\quad [\text{by } (3.5)]$$

$$= \int_C\ c(B)f\ dP = \int_{B\cap C}\ f\ dP$$

$$= \int_{B\cap C}\ E(f/\,\mathcal{B}\vee\mathcal{C}\,)\ dP\ .$$

SECTION 10. THE MARTINGALE THEOREM. The only thing along the-
se lines that remains to be discussed is the continuity of $E(f/\,\mathcal{C}\,)$
in $\mathcal{C}$. A typical and useful result is that if $\{\mathcal{C}_n\}$ is an increasing sequence of subfields of $\mathcal{S}$ (that is, $\mathcal{C}_n\subset\mathcal{C}_{n+1}$ for all n),
and if $\mathcal{C}$ is the subfield they generate, then

$$(10.1) \qquad E(f/\,\mathcal{C}_n\,)\ \longrightarrow\ E(f/\,\mathcal{C}\,)$$

19

P.R.Halmos

almost everywhere and in the mean. This is a non-trivial assertion; it is a special case of Doob's justly celebrated martingale convergence theorem. The special case $f = c(A)$ is all that we shall need; the assertion for that case is that

$$(10.2) \qquad P(A/\mathcal{C}_n) \;\rightarrow\; P(A/\mathcal{C})$$

almost everywhere. We shall give the proof for a statement of intermediate generality: we shall prove (10.1) in case f is bounded (and, of course, measurable). Mean convergence is then a consequence of the Lebesgue bounded convergence theorem.

Assume for a moment that it is already known that

$$(10.3) \qquad E(g/\mathcal{C}_n) \;\rightarrow\; g$$

almost everywhere whenever g is measurable $\mathcal{C}$ (and bounded). Note that in this case $E(g/\mathcal{C}) = g$. If f is an arbitrary bounded measurable function, write $g = E(f/\mathcal{C})$. It follows that g is bounded and measurable $\mathcal{C}$.

Since, by (8.1),

$$E(g/\mathcal{C}_n) = E(f/\mathcal{C}_n) \;,$$

the desired result (10.1) is proved. Conclusion: it is sufficient to prove (10.3).

If g is bounded and measurable $\mathcal{C}$, then it is the limit almost everywhere of a uniformly convergent sequence $\{g_k\}$ of simple functions measurable $\mathcal{C}$. Since, by (3.4),

$$|E(g/\mathcal{C}_n) - E(g_k/\mathcal{C}_n)| \leqq E(|g - g_k|/\mathcal{C}_n),$$

it follows that

P.R.Halmos

(10.4)
$$|E(g/\mathfrak{C}_n) - g| \lessgtr$$

$$E(|g - g_k|/\mathfrak{C}_n) + |E(g_k/\mathfrak{C}_n) - g_k| + |g_k - g|$$

The first and third terms on the right side of (10.4) tend to zero uniformly as $k \rightarrow \infty$, Conclusion: it is sufficient to prove (10.3) for simple functions measurable $\mathfrak{C}$, and hence for the characteristic functions of sets in $\mathfrak{C}$. The desired result has thus been reduced to

(10.5)
$$P(A/\mathfrak{C}_n) \rightarrow c(A)$$

almost everywhere, whenever A is in $\mathfrak{C}$.

Let ϵ and δ be arbitrary positive numbers less than 1. Since the union of the fields $\mathfrak{C}_n$ is dense in $\mathfrak{C}$, there exists a set B in that union, say B is in $\mathfrak{C}_k$, such that $P(A + B) < \epsilon \, \delta/2$ (where the plus sign denotes Boolean sum: $A + B = (A - B) \cup (B - A)$) Write

$$D_n = \{x: 1 - P(A/\mathfrak{C}_n)(x) \lessgtr \epsilon\} ,$$

and write $\{F_n\}$ for the sequence obtained by disjointing the sequence $\{B \cap D_n\}$; that is, $F_1 = B \cap D_1$,

$$F_2 = (B \cap D_2) - F_1 , \quad F_3 = (B \cap D_3) - (F_1 \cup F_2), \quad \text{etc.}$$

Observe that F_n is in $\mathfrak{C}_n$ whenever $n \gtreqless k$. It follows that if $n \gtreqless k$, then

$$P(F_n - A) = \int_{F_n} [1 - P(A/\mathfrak{C}_n)] \, dP \gtreqless P(F_n)\epsilon ,$$

and hence that if $F = \bigcup_{n \gtreqless k} F_n$, then

$$P(F - A) \gtreqless P(F)\epsilon .$$

P.R.Halmos

This implies that $P(B - A) \geqq P(F)\epsilon$. Since $P(B - A) \leqq P(B + A) <$ $< \epsilon\delta/2$, it follows that $P(F) < \delta/2$.

If x is in $B - F$, then $P(A/\mathcal{C}_n)(x) \geqq 1 - \epsilon$ whenever $n \geqq k$. This can be expressed as follows: if $n \geqq k$, then $P(A/\mathcal{C}_n) \geqq 1 - \epsilon$ throughout B, except possibly in a set of measure less than δ ; recall that $\epsilon\delta/2 + \delta/2 < \delta$. Since ϵ is arbitrarily small, it follows that $P(A/\mathcal{C}_n)$ converges to 1 throughout A, except possibly in a set of measure less than δ ; since δ is arbitrarily small, it follows that $P(A/\mathcal{C}_n)$ converges to 1 almost everywhere in A . This result applied to X - A in place of A shows that $P(A/\mathcal{C}_n)$ converges to $c(A)$ almost everywhere.

P.R.Halmos

CHAPTER II. INFORMATION

SECTION 11. MOTIVATION. What is a reasonable numerical measure of the amount of information conveyed by a statement? How much information, for instance, do we get about a point x of X when we are told that x belongs to a subset A of X? It seems reasonable to require that the answer should depend on the size of A (that is, on P(A)) and on nothing else. In other words, the answer should be expressed in terms of a function F on the unit interval; the amount of information conveyed by A shall be F(P(A)). Two further reasonable requirements are that the function F be non-negative and continuous.

Two experiments (or, alternatively, two statements) do not necessarily yield more information than one. If, however, two experiments are independent of each other, then it is reasonable to expect that the amount of information obtainable from the two together is the sum of the two separate amounts of information. If we perform two independent experiments, and if the result of the first tells us that x is in A and the result of the second tells us that x is in B , then we know that x belongs to a set (namely A $\cap$ B) of measure P(A)P(B) (by independence). The requirement of additivity implies, therefore, that the function F should satisfy the functional equation

$$F(st) = F(s) + F(t)$$

throughout its domain.

It is well known, and it is easy to prove, that the conditions imposed on F uniquely characterize F in the interval (0,1] , to wi-

P.R.Halmos

thin a multiplicative constant. Indeed, two inductions (one for the
numerator and one for the denominator) show that $F(t^r) = rF(t)$ whenever r is a positive rational number. Continuity yields the same
result whenever r is a positive real number. Put $t = \frac{1}{e}$ to get
$F(e^{-r}) = rF(e^{-1})$, or, in other words,

$$F(t) = F(e^{-1})(- \log t).$$

Conclusion: except for a constant factor, $F(t) = - \log t$; the a-
mount of information conveyed by the assertion that x is in A is
satisfactorily measured by $- \log P(A)$.

SECTION 12. DEFINITION. As the preceding discussion indicates,
the mathematical model of an experiment with a finite number of possible outcomes is a finite partition of X into measurable sets, or
equivalently, and from the point of view of the intended applications more elegantly, a finite subfield of $\mathcal{S}$. The amount of information conveyed by one of the possible outcomes of an experiment
is a number depending on that outcome; in other words, the amount
of information is a function of outcomes (associated with some experiments). These considerations motivate the following definition.
If $\mathcal{a}$ is any finite subfield of $\mathcal{S}$, the information function $I(\mathcal{a})$
associated with $\mathcal{a}$ is the function whose value at each point of each
atom A of $\mathcal{a}$ is $- \log P(A)$; equivalently

$$(12.1) \qquad I(\mathcal{a}) = - \sum_A c(A) \log P(A) .$$

Conditional information is a natural and useful generalization
of this concept; it is obtained by using conditional probabilities
instead of absolute probabilities. Explicitly, if $\mathcal{a}$ and $\mathcal{C}$ are subfields of $\mathcal{S}$, with $\mathcal{a}$ finite, then, by definition,

P.R.Halmos

$$(12.2) \qquad I(\mathcal{a}/\mathcal{C}) = - \Sigma_A \, c(A) \, \log P(A/\mathcal{C}) \, .$$

Observe that $I(\mathcal{a}/\mathcal{C})$ is always non-negative. The connection between conditional information and absolute information is that

$$(12.3) \qquad I(\mathcal{a}/2) = I(\mathcal{a}) \, ;$$

see (5.2). The function $I(\mathcal{a}/\mathcal{C})$ is measurable $\mathcal{a} \vee \mathcal{C}$, and, in particular, $I(\mathcal{a})$ is measurable $\mathcal{a}$.

SECTION 13. TRANSFORMATIONS. If T is a measure-preserving transformation on X , then

$$(13.1) \qquad I(\mathcal{a}/\mathcal{C})T = I(T^{-1}\mathcal{a}/T^{-1}\mathcal{C}) \, ;$$

the proof is a straightforward application of (7.2). It follows in particular, that

$$(13.2) \qquad I(\mathcal{a})T - I(T^{-1}\mathcal{a})$$

SECTION 14. INFORMATION ZERO. If $\mathcal{a} \subset \mathcal{C}$, then

$$(14.1) \qquad I(\mathcal{a}/\mathcal{C}) = 0 \, .$$

The proof is based on (5.1) and the convention that if $t = 0$, then $\log t = 0$. It follows in particular, with $\mathcal{a} = 2$ and $\mathcal{C} = 2$, that

$$(14.2) \qquad I(2) = 0 \, .$$

These equations are in harmony with the intuitive meaning of information. Thus, for example, (14.1) expresses that if before some particular experiment is performed more is already known than the experiment can possibly reveal, then the amount of information conveyed by the outcome is certainly zero.

P.R.Halmos

The converse of the conclusion (14.1) is true; the assertion is that if $I(\mathcal{A}/\mathcal{C}) = 0$ almost everywhere, then

$$(14.3) \qquad \mathcal{A} \subset \mathcal{C} .$$

Indeed, if $I(\mathcal{A}/\mathcal{C}) = 0$ almost everywhere, then $c(A) \log P(A/\mathcal{C}) = 0$ almost everywhere for each atom A of $\mathcal{A}$. This implies that $P(A/\mathcal{C})$ is a characteristic function and hence (by (5.3)) that A is in $\mathcal{C}$.

SECTION 15. ADDITIVITY. Conditional probability (of a set with respect to a field) is algebraically well-behaved as a function of its first argument and analytically well-behaved as a function of its second argument. These facts are reflected by the behavior of conditional information.

If $\mathcal{A}$, $\mathcal{B}$ and $\mathcal{C}$ are fields such that $\mathcal{A}$ and $\mathcal{B}$ are finite, then

$$(15.1) \qquad I(\mathcal{A} \vee \mathcal{B} / \mathcal{C}) = I(\mathcal{B}/\mathcal{C}) + I(\mathcal{A}/ \mathcal{B} \vee \mathcal{C}) .$$

The left side of (15.1) is symmetric in $\mathcal{A}$ and $\mathcal{B}$, whereas the right side is apparently not. This is no cause for alarm; (15.1) is merely one of two equations, which between them describe the whole (symmetric) truth. If $\mathcal{C} = 2$, then (15.1) becomes

$$(15.2) \qquad I(\mathcal{A} \vee \mathcal{B}) = I(\mathcal{B}) + I(\mathcal{A}/\mathcal{B}) ;$$

if $\mathcal{A} \subset \mathcal{C}$, then (15.1) becomes

$$(15.3) \qquad I(\mathcal{A} \vee \mathcal{B} / \mathcal{C}) = I(\mathcal{B}/\mathcal{C}) .$$

If, in particular, $\mathcal{C} = \mathcal{A}$, then

$$(15.4) \qquad I(\mathcal{A} \vee \mathcal{B} / \mathcal{A}) = I(\mathcal{B}/\mathcal{A}) .$$

P.R.Halmos

If, finally, $\mathcal{A} \subset \mathcal{B}$, then

$$(15.5) \qquad I(\mathcal{A}/\mathcal{C}) \leqq I(\mathcal{B}/\mathcal{C}) \ ;$$

this follows from (15.1) and the fact that $I(\mathcal{B}/\mathcal{A}\vee\mathcal{C})$ is non-negative.

The proof of (15.1) is the following computation:

$$
\begin{aligned}
I(\mathcal{A}\vee\mathcal{B}/\mathcal{C}) &= - \Sigma_A \Sigma_B \ c(A)c(B) \ \log P(A \cap B/\mathcal{C}) \\[2mm]
&= - \Sigma_A \Sigma_B \ c(A)c(B) \ \log \left(P(B/\mathcal{C}) \ \frac{P(A \cap B/\mathcal{C})}{P(B/\mathcal{C})} \right) \\[2mm]
&= - \Sigma_A \Sigma_B \ c(A)c(B) \ \log P(B/\mathcal{C}) - \Sigma_A \Sigma_B \ c(A)c(B) \ \log \frac{P(A \cap B/\mathcal{C})}{P(B/\mathcal{C})} \\[2mm]
&= - \Sigma_A \ c(A) \ \Sigma_B \ \log P(B/\mathcal{C}) - \Sigma_A \ c(B) \ \frac{P(A \cap B/\mathcal{C})}{P(B/\mathcal{C})} \\[2mm]
&= \Sigma_A \ c(A) \ I(\mathcal{B}/\mathcal{C}) - \Sigma_A \ c(A) \ \log P(A/\mathcal{B}\vee\mathcal{C}) \\[2mm]
&= I(\mathcal{B}/\mathcal{C}) + I(\mathcal{A}/\mathcal{B}\vee\mathcal{C}) \ .
\end{aligned}
$$

SECTION 16. FINITE ADDITIVITY. The following assertion is a useful corollary of (15.3). If $\mathcal{B}_0$, $\mathcal{B}_1, \ldots, \mathcal{B}_n$ are finite fields (with $\mathcal{B}_0 = 2$, and $n = 1, 2, 3, \ldots$), then

$$(16.1) \qquad I\left(\bigvee_{i=1}^{n} \mathcal{B}_i \right) = \Sigma_{k=1}^{n} \ I\left(\mathcal{B}_k / \bigvee_{i=0}^{k-1} \mathcal{B}_i \right) \ .$$

The proof is inductive. For $n = 1$, the assertion reduces to $I(\mathcal{B}_1) = I(\mathcal{B}_1/2)$. For the induction step, use (15.3) as follows:

P.R.Halmos

$$I(\bigvee_{i=1}^{n+1} \mathcal{B}_i) = I(\bigvee_{i=1}^{n} \mathcal{B}_i \vee \mathcal{B}_{n+1})$$

$$= I(\bigvee_{i=1}^{n} \mathcal{B}_i) + I(\mathcal{B}_{n+1} / \bigvee_{i=1}^{n} \mathcal{B}_i) .$$

An important special case of (16.1) is obtained as follows. Suppose that $\mathcal{a}$ is a finite subfield of $\mathcal{b}$ and that T is a measure-preserving transformation on X . For each positive integer n , write $\mathcal{B}_i = T^{-(n-i)} \mathcal{a}$ $(i = 1,\ldots,n)$, and apply (16.1). The result is

$$I(\bigvee_{i=1}^{n} T^{-(n-i)} \mathcal{a}) = I(T^{-(n-1)} \mathcal{a})$$

$$+ \sum_{k=2}^{n} I(T^{-(n-k)} \mathcal{a} / \bigvee_{i=1}^{k-1} T^{-(n-i)} \mathcal{a})$$

for $n > 1$. This can be rewritten in the form

$$I(\bigvee_{i=0}^{n-1} T^{-i} \mathcal{a}) = I(T^{-(n-1)} \mathcal{a})$$

$$+ \sum_{k=1}^{n-1} I(T^{-(n-k-1)} \mathcal{a} / \bigvee_{i=1}^{k} T^{-(n-i)} \mathcal{a}).$$

Since $T^{-(n-i)} \mathcal{a} = T^{-(n-k-1)} T^{-(k-i+1)} \mathcal{a}$ it follows that

$$I(\bigvee_{i=0}^{n-1} T^{-i} \mathcal{a}) = I(\mathcal{a})T^{n-1}$$

$$+ \sum_{k=1}^{n-1} I(\mathcal{a} / \bigvee_{i=1}^{k} T^{-(k-i+1)} \mathcal{a}) T^{n-k-1}$$

or, equivalently,

$$(16.2) \qquad I(\bigvee_{i=0}^{n-1} T^{-i} \mathcal{a}) = I(\mathcal{a})T^{n-1}$$

$$+ \sum_{k=1}^{n-1} I(\mathcal{a} / \bigvee_{i=1}^{k} T^{-i} \mathcal{a})T^{n-k-1}$$

whenever $n > 1$.

[The preceding computation made use of the fact that

P.R.Halmos

$$(16.3) \qquad T^{-1}(\bigvee \mathcal{E}) = \bigvee (T^{-1} \mathcal{E})$$

for any collection $\mathcal{E}$ of sets; $\bigvee$ here denotes the generated field. To prove this, observe that $\mathcal{E} \subset \bigvee \mathcal{E}$, so that $T^{-1} \mathcal{E} \subset T^{-1}(\bigvee \mathcal{E})$; since $T^{-1} (\bigvee \mathcal{E})$ is a field, it follows that

$$\bigvee (T^{-1} \mathcal{E}) \subset T^{-1}(\bigvee \mathcal{E}) .$$

For the reverse inclusion observe that $T^{-1} \mathcal{E} \subset \bigvee (T^{-1} \mathcal{E})$.

Since the collection of those sets A for which $T^{-1} A$ belongs to $\bigvee (T^{-1} \mathcal{E})$ is a field, it follows that

$$T^{-1}(\bigvee \mathcal{E}) \subset \bigvee (T^{-1} \mathcal{E}) ,$$

and the proof is complete.

If, in particular, $\mathcal{B}$ and $\mathcal{C}$ are fields, then put $\mathcal{E} = \mathcal{B} \cup \mathcal{C}$ and apply (16.3); the result is that

$$T^{-1}(\mathcal{A} \vee \mathcal{B}) = T^{-1} \mathcal{A} \vee T^{-1} \mathcal{B} .$$

[The similar facts about infinite suprema are proved tha same way].

SECTION 17. CONVERGENCE. If $\{ \mathcal{C}_n \}$ is an increasing sequence of subfields of $\mathcal{B}$, and if $\mathcal{C} = \bigvee_n \mathcal{C}_n$, then

$$(17.1) \qquad I(\mathcal{A} / \mathcal{C}_n) \to I(\mathcal{A} / \mathcal{C})$$

almost everywhere. The proof of this assertion is immediate from the definition (12.2) and the convergence theorem (10.2).

For some purposes convergence in the mean is more useful than almost everywhere convergence. Convergence in the mean is closely connected with uniform integrability. Recall that a se-

P.R.Halmos

quence $\{f_n\}$ of measurable functions on X is uniformly integrable
if

$$\int_{\{x:\ |f_n(x)| \geqq t\}} |f_n|\ dP$$

tends to 0 , as $t \to \infty$, uniformly in n . The facts are these. If
f is an integrable function and if $\{f_n\}$ is a sequence of integrable functions such that $f_n \to f$ in the mean, then $\{f_n\}$ is uniformly integrable. If, conversely, f is a measurable function, $\{f_n\}$
is a uniformly integrable sequence, and if $f_n \to f$ almost everywhere, then f is integrable and $f_n \to f$ in the mean. In view of these
facts, the way to prove that (17.1) may also be interpreted in the
sense of mean convergence is to prove uniform integrability. Note
that even the integrability of one $I(\mathcal{A}/\mathcal{C})$ needs proof; there is
no obvious boundedness to invoke.

UNIFORM INTEGRABILITY THEOREM. If N is a positive integer,
then the collection of all functions of the form $I(\mathcal{A}/\mathcal{C})$ (where
$\mathcal{A}$ is a finite subfield of $\mathcal{S}$ with not more than N atoms and $\mathcal{C}$
is an arbitrary subfield of $\mathcal{S}$) is uniformly integrable.

PROOF. Take N, $\mathcal{A}$, and $\mathcal{C}$ as described, and let r and s be
positive numbers, $r \leqq s$. Write

$$D = \{x:\ r \leqq I(\mathcal{A}/\mathcal{C})(x) \leqq s\ \} \ .$$

If A is an atom of $\mathcal{A}$, then $I(\mathcal{A}/\mathcal{C}) = -\log P(A/\mathcal{C})$ on A . It
follows that if

$$C_A = \{x:\ r \leqq -\log P(A/\mathcal{C})(x) \leqq s\ \} \ ,$$

then $A \cap D = A \cap C_A$. Since $P(A/\mathcal{C})$ is measurable $\mathcal{C}$, the set C_A
belongs to $\mathcal{C}$ for each A . Since on C_A

$$\log P(A/\mathcal{C}) \leqq -r$$

P.R.Halmos

or

$$P(A/\mathcal{C}) \leqq e^{-r},$$

it follows that

$$P(A \cap D) = P(A \cap C_A) = \int_{C_A} P(A/\mathcal{C})\, dP$$

$$\leqq e^{-r} P(C_A) \leqq e^{-r}.$$

Since $I(\mathcal{a}/\mathcal{C}) \leqq s$ on D,

$$\int_{A \cap D} I(\mathcal{a}/\mathcal{C})\, dP \leqq sP(A \cap D) \leqq se^{-r}.$$

Sum over all the atoms A of $\mathcal{a}$ to get

$$\int_D I(\mathcal{a}/\mathcal{C})\, dP \leqq Nse^{r}.$$

Put $r = t + n$, $s = t + n - 1$ $(n = 0, 1, 2, \ldots)$; it follows that

$$(17.2) \qquad \int_{\{x:\ I(\mathcal{a}/\mathcal{C})(x)\, \geqq\, t\}} I(\mathcal{a}/\mathcal{C})\, dP \leqq N \sum_{n=0}^{\infty} \frac{t+n+1}{e^{t+n}}.$$

Since the sum in (17.2) tends to 0 as $t \to \infty$, the proof of
uniform integrability is complete.

COROLLARY. If $\{\mathcal{C}_n\}$ is an increasing sequence of subfields
of $\mathcal{S}$ with $\bigvee_n \mathcal{C}_n = \mathcal{C}$, then $I(\mathcal{a}/\mathcal{C}_n) \to I(\mathcal{a}/\mathcal{C})$ in the mean
for each finite field $\mathcal{a}$.

SECTION 18. MCMILLAN'S THEOREM. The preceding corollary can
be applied to the following important situation. Suppose that $\mathcal{a}$
is a finite subfield of $\mathcal{S}$ and that T is a measure-preserving transformation on X. Write

P.R.Halmos

$$f_n = \bar{I}\left(\bigvee_{i=0}^{n-1} T^{-i} a \right), \qquad n = 1, 2, 3,\ldots,$$

$$g_0 = I(a),$$

$$g_n = I\left(a / \bigvee_{i=1}^{n} T^{-i} a \right), \qquad n = 1, 2, 3,\ldots\ .$$

It follows from (16.2) that

$$f_n = \sum_{k=0}^{n-1} g_k T^{n-k-1}, \qquad n = 1, 2, 3,\ldots\ .$$

If $g = I\left(a / \bigvee_{i=1}^{\infty} T^{-i} a \right)$, then (17.1) implies that

$g_n \to g$ almost everywhere, and the corollary of the uniform integrability theorem implies that $g_n \to g$ in the mean. The ergodic
theorem applied to g implies that the averages

$$\frac{1}{n} \sum_{k=0}^{n-1} g T^{n-k-1}$$

tend to some invariant integrable function h both almost everywhere and in the mean. From these facts it is easy to deduce the
following assertion, known as McMillan's theorem: $\frac{1}{n} f_n \to h$ in
the mean. Indeed, if $||f|| = \int |f|\, dP$, then

$$\left\| \frac{1}{n} f_n - h \right\| = \left\| \frac{1}{n} \sum_{k=0}^{n-1} g_k T^{n-k-1} - h \right\|$$

$$\leq \left\| \frac{1}{n} \sum_{k=0}^{n-1} (g_k T^{n-k-1} - g T^{n-k-1}) \right\| + \left\| \frac{1}{n} \sum_{k=0}^{n-1} g T^{n-k-1} - h \right\|$$

$$\leq \frac{1}{n} \sum_{k=0}^{n-1} \| g_k - g \| + \left\| \frac{1}{n} \sum_{k=0}^{n-1} g T^{k} - h \right\|,$$

and the proof of McMillan's theorem (for not necessarily invertible transformations) is complete. Presumably McMillan's theorem can
be sharpened by adding the conclusion that $\frac{1}{n} f_n \to g$ almost everywhere.

P.R.Halmos

CHAPTER III. ENTROPY

SECTION 19. DEFINITION. What we have been studying so far is the amount of information (absolute and conditional) furnished by an experiment. An associated concept of importance is the average amount of information so furnished. The best sense of "average" here is "expected value", where the expectation in question may itself be absolute or conditional. Suppose, to be specific, that $\mathcal{A}$, $\mathcal{C}$ and $\mathcal{D}$ are subfields of $\mathcal{S}$, with $\mathcal{A}$ finite, and write

$$H((\mathcal{A}/\mathcal{C})/\mathcal{D}) = E(I(\mathcal{A}/\mathcal{C})/\mathcal{D}) ;$$

the function H is called entropy, the entropy of $\mathcal{A}$, the conditional entropy of $\mathcal{A}$ with respect to $\mathcal{C}$, or the mean conditional entropy with respect to $\mathcal{D}$ of $\mathcal{A}$ with respect to $\mathcal{C}$.

If $\mathcal{D} = 2$, then the mean conditional entropy with respect to $\mathcal{D}$ is a constant $\bar{H}(\mathcal{A}/\mathcal{C})$; it follows from the elementary properties of conditional expectations that

$$(19.1) \qquad \bar{H}(\mathcal{A}/\mathcal{C}) = \int I(\mathcal{A}/\mathcal{C}) \, dP .$$

If $\mathcal{D} = \mathcal{C}$, the mean conditional entropy with respect to $\mathcal{D}$ is a function $H(\mathcal{A}/\mathcal{C})$,

$$(19.2) \qquad H(\mathcal{A}/\mathcal{C}) = E(I(\mathcal{A}/\mathcal{C})/\mathcal{C});$$

it follows from the elementary properties of conditional expectations that

$$(19.3) \qquad \int H(\mathcal{A}/\mathcal{C}) \, dP = \bar{H}(\mathcal{A}/\mathcal{C}) .$$

If $H(\mathcal{A}) = H(\mathcal{A}/2)$ and $\bar{H}(\mathcal{A}) = \bar{H}(\mathcal{A}/2)$, then it follows from

P.R.Halmos

(19.2) that $H(\mathcal{A})$ is a constant, and it follows from (19.3) that the constant is $\hat{H}(\mathcal{A})$. Note also that

$$(19.4) \qquad H(\mathcal{A}) = \bar{H}(\mathcal{A}) = \int I(\mathcal{A})\, dP\ .$$

Conditional entropy can be computed directly from the definition (12.2) of information;

$$(19.5) \qquad H(\mathcal{A}/\mathcal{B}) = -\sum_A P(A/\mathcal{B})\, \log P(A/\mathcal{B})\ .$$

If, in particular, $\mathcal{B} = 2$, then

$$(19.6) \qquad \bar{H}(\mathcal{A}) = -\sum_A P(A)\, \log P(A)\ .$$

SECTION 20. TRANSFORMATIONS. Most of the properties of entropies are relatively easy to derive from the corresponding properties of information. Thus, for instance, if T is a measure-preserving transformation on X , then, by (13.1) and (7.1),

$$H(\mathcal{A}/\mathcal{B})T = H(T^{-1}\mathcal{A}/T^{-1}\mathcal{B})\ ,$$

and, in particular,

$$\bar{H}(\mathcal{A}/\mathcal{B}) = \bar{H}(T^{-1}\mathcal{A}/T^{-1}\mathcal{B})$$

and

$$\bar{H}(\mathcal{A}) = \bar{H}(T^{-1}\mathcal{A})\ .$$

SECTION 21. ENTROPY ZERO. If $\mathcal{A} \subset \mathcal{B}$, then

$$(21.1) \qquad H(\mathcal{A}/\mathcal{B}) = 0\ ,$$

and, in particular,

P.R.Halmos

$$\bar{H}(\mathcal{a}/\mathcal{C}) = 0$$

(21.2)

and

$$\bar{H}(2) = 0 ;$$

cf. (14.1) and (14.2).

Conversely if $H(\mathcal{a}/\mathcal{C}) = 0$ almost everywhere (or, equivalently, if $\bar{H}(\mathcal{a}/\mathcal{C}) = 0$), then $I(\mathcal{a}/\mathcal{C}) = 0$ almost everywhere (by (19.1)), and therefore $\mathcal{a} \subset \mathcal{C}$ (by (14.3)).

SECTION 22. CONCAVITY. If $\mathcal{a}$, $\mathcal{B}$ and $\mathcal{C}$ are fields such that $\mathcal{a}$ is finite and $\mathcal{B} \subset \mathcal{C}$, then

$$\bar{H}(\mathcal{a}/\mathcal{C}) \leqq \bar{H}(\mathcal{a}/\mathcal{B}) .$$

(22.1)

The proof is an application of Jensen's inequality. If $F(t) = t \log t$ whenever $t > 0$ and $F(0) = 0$, then F is a continuous convex function; it follows that

$$F(E(P(A/\mathcal{C})/\mathcal{B}) \leqq E(F(P(A/\mathcal{C})/\mathcal{B})$$

(22.2)

for each atom A of $\mathcal{a}$. By (8.3) the left side of (22.2) is equal to $F(P(A/\mathcal{B}))$; this implies that

$$\Sigma_A P(A/\mathcal{B}) \log P(A/\mathcal{B}) \leqq E(\Sigma_A P(A/\mathcal{C}) \log P(A/\mathcal{C})/\mathcal{B}) .$$

To get (22.1), change sign and integrate; cf. (19.5).

The special case $\mathcal{B} = 2$ is worthy of note :

$$\bar{H}(\mathcal{a}/\mathcal{C}) \leqq \bar{H}(\mathcal{a}) .$$

(22.3)

P.R.Halmos

SECTION 23. ADDITIVITY. If $\mathcal{A}$, $\mathcal{B}$, and $\mathcal{C}$ are fields such that $\mathcal{A}$ and $\mathcal{B}$ are finite, then, by (15.1),

$$(23.1) \qquad H(\mathcal{A} \vee \mathcal{B} / \mathcal{C}) = H(\mathcal{B}/\mathcal{C}) + H((\mathcal{A}/\mathcal{B}\vee\mathcal{C})/\mathcal{C}) .$$

It follows that

$$(23.2) \qquad \bar{H}(\mathcal{A} \vee \mathcal{B} / \mathcal{C}) = \bar{H}(\mathcal{B}/\mathcal{C}) + \bar{H}(\mathcal{A}/\mathcal{B}\vee\mathcal{C}) ,$$

and, in particular,

$$(23.4) \qquad \bar{H}(\mathcal{A} \vee \mathcal{B}) = \bar{H}(\mathcal{B}) + \bar{H}(\mathcal{A}/\mathcal{B}) .$$

Further special cases: if $\mathcal{A} \subset \mathcal{C}$, then

$$H(\mathcal{A} \vee \mathcal{B} / \mathcal{C}) = H(\mathcal{B}/\mathcal{C}) ,$$

$$\bar{H}(\mathcal{A} \vee \mathcal{B} / \mathcal{C}) = \bar{H}(\mathcal{B}/\mathcal{C}) ,$$

$$\bar{H}(\mathcal{A} \vee \mathcal{B} / \mathcal{A}) = \bar{H}(\mathcal{B}/\mathcal{A}) .$$

It follows from (23.2) and (22.1) that

$$(23.5) \qquad \bar{H}(\mathcal{A} \vee \mathcal{B} / \mathcal{C}) \leqq \bar{H}(\mathcal{A}/\mathcal{C}) + \bar{H}(\mathcal{B}/\mathcal{C}) ,$$

and hence, in particular (cf. (23.4)),

$$(23.6) \qquad \bar{H}(\mathcal{A} \vee \mathcal{B}) \leqq \bar{H}(\mathcal{A}) + \bar{H}(\mathcal{B}) .$$

If $\mathcal{A} \subset \mathcal{B}$, then (by (15.5))

$$H(\mathcal{A}/\mathcal{C}) \leqq H(\mathcal{B}/\mathcal{C})$$

and therefore,

$$\bar{H}(\mathcal{A}/\mathcal{C}) \leqq \bar{H}(\mathcal{B}/\mathcal{C}) .$$

P.R.Halmos

It follows (put $\mathcal{C} = 2$) that if $\mathcal{a} \subset \mathcal{B}$, then

$$\bar{H}(\mathcal{a}) \leqq \bar{H}(\mathcal{B}) \ .$$

SECTION 24. FINITE ADDITIVITY. If $\mathcal{B}_0$, $\mathcal{B}_1, \ldots, \mathcal{B}_n$ are finite fields (with $\mathcal{B}_0 = 2$, and $n = 1, 2, 3, \ldots$), then

$$(24.1) \qquad \bar{H}\left(\bigvee_{i=1}^{n} \mathcal{B}_i \right) = \sum_{k=1}^{n} \bar{H}\left(\mathcal{B}_k \Big/ \bigvee_{i=0}^{k-1} \mathcal{B}_i \right) ;$$

this is an immediate consequence of (16.4). If T is a measure-pre-serving transformation on X , then (16.2) applies; the conclusion is that

$$(24.2) \qquad \bar{H}\left(\bigvee_{i=0}^{n-1} T^{-i} \mathcal{a} \right) = \bar{H}(\mathcal{a}) + \sum_{k=1}^{n-1} \bar{H}\left(\mathcal{a} \Big/ \bigvee_{i=1}^{k} T^{-i} \mathcal{a} \right)$$

whenever $n > 1$.

SECTION 25. CONVERGENCE. If $\{ \mathcal{C}_n \}$ is an increasing sequence of subfields of $\mathcal{S}$, and if $\mathcal{C} = \bigvee_n \mathcal{C}_n$, then

$$(25.1) \qquad H(\mathcal{a} / \mathcal{C}_n) \to H(\mathcal{a} / \mathcal{C})$$

almost everywhere and in the mean. The proof is immediate from the convergence theorem (10.2) and the equation (19.5); recall that since t log t is bounded for t in $[0,1]$, the Lebesgue bounded convergence theorem is applicable. An immediate corollary is that

$$(25.2) \qquad \bar{H}(\mathcal{a} / \mathcal{C}_n) \to \bar{H}(\mathcal{a} / \mathcal{C}) \ .$$

P.R.Halmos

CHAPTER IV. APPLICATION

SECTION 26. RELATIVE ENTROPY. If T is a measure-preserving transformation on X , then

$$(26.1) \qquad \bar{H}(\alpha/ \bigvee_{i=1}^{k} T^{-i}\alpha) \to \bar{H}(\alpha/ \bigvee_{i=1}^{\infty} T^{-i}\alpha) ;$$

alternatively, the same conclusion can be derived by applying term-by-term integration to McMillan's theorem.

The limit in (26.2) is an important one in the theory of measure-preserving transformations. If we write

$$(26.3) \qquad h(\alpha, T) = \bar{H}(\alpha/ \bigvee_{i=1}^{\infty} T^{-i}\alpha) ,$$

then

$$(26.4) \qquad h(\alpha, T) = \lim \frac{1}{n} \bar{H}(\bigvee_{i=0}^{n-1} T^{-i}\alpha) .$$

We may call $h(\alpha, T)$ the entropy of T relative to α .

SECTION 27. ELEMENTARY PROPERTIES. If α and $\mathcal{B}$ are finite fields, then, by (23.2) and (26.4),

$$(27.1) \qquad h(\alpha \vee \mathcal{B}, T) \leqq h(\alpha, T) + h(\mathcal{B}, T) .$$

If $\alpha \subset \mathcal{B}$, then

$$(27.2) \qquad h(\alpha, T) \leqq h(\mathcal{B}, T) .$$

If S and T are commutative measure-preserving transformations, then

$$h(S^{-1}\alpha , T) = h(\alpha, T) .$$

Observe, indeed, that

P.R.Halmos

$$\bigvee_{i=1}^{\infty} T^{-i}(S^{-1}\alpha) = \bigvee_{i=1}^{\infty} S^{-1}(T^{-i}\alpha) = S^{-1}(\bigvee_{i=1}^{\infty} T^{-i}\alpha) \ ;$$

it follows that

$$\overline{H}(S^{-1}\alpha / \bigvee_{i=1}^{\infty} T^{-i}(S^{-1}\alpha))$$

$$= \overline{H}(S^{-1}\alpha / S^{-1}\bigvee_{i=1}^{\infty} T^{-i}\alpha) = \overline{H}(\alpha / \bigvee_{i=1}^{\infty} T^{-i}\alpha) \ .$$

A special case of value is

$$h(T^{-1}\alpha, \ T) = h(\alpha, T) \ .$$

SECTION 28. STRONG MONOTONY. The next result is that if T is invertible, then (27.2) can be given an infinitely sharper form. If, to be precise, α and $\mathcal{B}$ are finite fields such that

$$\alpha \subset \bigvee_{i=-\infty}^{+\infty} T^{i}\mathcal{B} \ ,$$

then

$$(28.1) \qquad\qquad h(\alpha, \ T) \leqq h(\mathcal{B}, \ T) \ .$$

The proof begins with the observation that

$$\bigvee_{i=0}^{k-1} T^{-i}\alpha \subset \bigvee_{i=0}^{k-1} T^{-i}\alpha \ \vee \ \bigvee_{j=-n-k+1}^{n} T^{j}\mathcal{B}$$

for all positive integers k and n . It follows that

$$(28.2) \qquad \overline{H}(\bigvee_{i=0}^{k-1} T^{-i}\alpha \ ) \ \leqq \ \overline{H}(\bigvee_{j=-n-k+1}^{+n} T^{j}\mathcal{B})$$

$$+ \ \overline{H}(\bigvee_{i=0}^{k-1} T^{-i}\alpha / \bigvee_{j=-n-k+1}^{+n} T^{j}\mathcal{B}) \ .$$

The second summand on the right side of (23.2) is dominated by

P.R.Halmos

$$\sum_{i=0}^{k-1} \bar{H}(T^{-i}\mathcal{A}\,/\,T^{-i}\bigvee_{j=-n-k+1}^{+n} T^{j+i}\mathcal{B}\,)$$

$$\leq \sum_{i=0}^{k-1} \bar{H}(\mathcal{A}\,/\bigvee_{j=-n-i}^{+n-i} T^{j+i}\mathcal{B}\,)$$

$$= \sum_{i=0}^{k-1} \bar{H}(\mathcal{A}\,/\bigvee_{j=-n}^{+n} T^{j}\mathcal{B}\,)\,,$$

and therefore

$$\bar{H}(\bigvee_{i=0}^{k-1} T^{-i}\mathcal{A}\,) \leq \bar{H}(T^{n}\bigvee_{j=0}^{2n+k-1} T^{-j}\mathcal{B}\,)$$

$$+ k\bar{H}(\mathcal{A}\,/\bigvee_{j=-n}^{+n} T^{j}\mathcal{B}\,)\,.$$

This implies that

(28.3)
$$\frac{1}{k}\,\bar{H}(\bigvee_{i=0}^{k-1} T^{-i}\mathcal{A}\,) \leq$$

$$\frac{2n+k}{k}\,\cdot\,\frac{1}{2n+k}\,\bar{H}(\bigvee_{j=0}^{2n+k-1} T^{-j}\mathcal{B}\,)$$

$$+ \bar{H}(\mathcal{A}\,/\bigvee_{j=-n}^{+n} T^{j}\mathcal{B}\,)\,.$$

Since $\mathcal{A} \subset \bigvee_{j=-\infty}^{+\infty} T^{j}\mathcal{B}$, it follows from (25.2) and (21.2) that the second summand on the right side of (28.3) tends to 0 as $n \to \infty$. Choose n large and then, for fixed n , let $k \to \infty$; the result is that $h(\mathcal{A},\,T)$ is dominated by $h(\mathcal{B},\,T_-)$ plus an arbitrarily small positive number. This completes the proof of (28.1).

SECTION 29. ALGEBRAIC PROPERTIES. The preceding results give some idea of how $h(\mathcal{A},\,T)$ depends on $\mathcal{A}$; we turn now to study the way it depends on T . The principal result along these lines is that

(29.1)
$$h(\mathcal{A},\,T^{k}) \leq k \cdot h(\mathcal{A},\,T)$$

P.R.Halmos

for each positive integer k .

To prove this, write $\mathcal{B} = \bigvee_{i=0}^{k-1} T^{-i} \mathcal{A}$; then

$$\bigvee_{j=0}^{n-1} (T^k)^{-j} \mathcal{B} = \bigvee_{j=0}^{n-1} T^{-jk} \bigvee_{i=0}^{k-1} T^{-i} \mathcal{A} = \bigvee_{i=0}^{kn-1} T^{-i} \mathcal{A} .$$

It follows that

$$\frac{1}{n} \bar{H}(\bigvee_{j=0}^{n-1} (T^k)^{-j} \mathcal{B}) = k \cdot \frac{1}{kn} \bar{H}(\bigvee_{i=0}^{kn-1} T^{-i} \mathcal{A})$$

and hence that

$$(29.2) \qquad h(\mathcal{B}, T^k) = k \cdot h(\mathcal{A}, T).$$

Since $\mathcal{A} \subset \mathcal{B}$, the assertion (29.1) follows from (27.2) (for T^k).

If T is invertible, then

$$(29.3) \qquad h(\mathcal{A}, T^{-1}) = h(\mathcal{A}, T) .$$

This follows immediately from the equations

$$\bar{H}(\bigvee_{i=0}^{n-1} T^i \mathcal{A}) = \bar{H}(T^{n-1} \bigvee_{i=0}^{n-1} T^{-i} \mathcal{A}) = \bar{H}(\bigvee_{i=0}^{n-1} T^{-i} \mathcal{A}) .$$

Combining (29.1) and (29.3), we obtain

$$h(\mathcal{A}, T^k) \leqq |k| \cdot h(\mathcal{A}, T)$$

for every integer k .

SECTION 30. ENTROPY. The entropy $h(T)$ of a measure-preserving transformation T is defined by

$$(30.1) \qquad h(T) = \sup h(\mathcal{A}, T) ,$$

where the supremum extends over all finite subfields of $\mathcal{S}$.

Since, by (29.2).,

P.R.Halmos

$$k \cdot h(\mathcal{a}, T) = h\left(\bigvee_{i=0}^{k-1} T^{-i}\mathcal{a}\, ,\, T^k\right) \leqq h(T^k) \, ,$$

and since, by (29.1),

$$h(\mathcal{a},\, T^k) \leqq k \cdot h(\mathcal{a}, T) \leqq k \cdot h(T) \, ,$$

it follows that

$$(30.2) \qquad h(T^k) = k \cdot h(T)$$

for every positive integer k .

If T is invertible, then, by (29.3),

$$h(T^{-1}) = h(T) \, ,$$

and therefore

$$h(T^k) = |k| \cdot h(T)$$

for every integer k .

SECTION 31. GENERATED FIELDS. The following result is an efficient tool for calculating the entropy of some transformations. If T is invertible and if $\mathcal{B}$ is a finite subfield of $\mathcal{S}$ such that

$$\bigvee_{i=-\infty}^{+\infty} T^i \mathcal{B} = \mathcal{S} \, ,$$

then

$$(31.1) \qquad h(T) = h(\mathcal{B}, T) \, .$$

Indeed, (28.1) implies that

$$h(\mathcal{a},\, T) \leqq h(\mathcal{B},\, T)$$

for all finite subfields $\mathcal{a}$ of $\mathcal{S}$, and hence that

P.R.Halmos

$$h(T) \leqq h(\mathcal{B}, T) .$$

The reverse inequality follows from the definition of $h(T)$.

A curious corollary of the preceding result is the assertion that if T is invertible and if $\mathcal{B}$ is a finite subfield of $\mathcal{S}$ such that

$$\bigvee_{i=0}^{\infty} T^{-i} \mathcal{B} = \mathcal{S}$$

then

$$h(T) = 0 .$$

Indeed, (31.1) implies that $h(T) = h(\mathcal{B}, T)$. The invertibility of T implies that $T^{-1}\mathcal{S} = \mathcal{S}$ and therefore, since

$$T^{-1} \mathcal{S} = \bigvee_{i=1}^{\infty} T^{-i} \mathcal{B} ,$$

that $\mathcal{B} \subset \bigvee_{i=1}^{\infty} T^{-i} \mathcal{B}$. The conclusion now follows from (26.3) and (21.2).

SECTION 32. EXAMPLES. We are now in a position to compute the entropy of some transformations.

Suppose, to begin with, that T is the identity. In that case $T^{-i}\mathcal{a} = \mathcal{a}$ for all $\mathcal{a}$, so that $\bigvee_{i=0}^{n-1} T^{-i} \mathcal{a} = \mathcal{a}$ for all $\mathcal{a}$ and all n . This implies that $h(\mathcal{a}, T) = \lim \frac{1}{n} \bar{H}(\mathcal{a}) = 0$ for all $\mathcal{a}$, and hence that $h(T) = 0$.

If T has finite order, i.e., T^k is the identity for some positive integer k , then $h(T^k) = 0$ by the preceding paragraph, and it follows (from (30.2)) that $h(T) = 0$.

Suppose next that X is the circle group and that T is a rotation, $Tx = cx$, where c is not a root of unity. Let $\mathcal{B}$ be the field consisting of the half-open top arc and its complement (and,

P.R.Halmos

of course, the empty set and X). For suitable values of i the result of applying T^{-i} to the top arc yields arbitrarily small arcs, which can then be rotated around to nearly arbitrary positions. (All rotations needed are by T^{-i} , i = 0, 1, 2,...). It follows that every half-open arc is approximable arbitrarily closely by sets in $\bigvee_{i=0}^{\infty} T^{-i}\mathcal{B}$, and hence that every such arc belongs to $\bigvee_{i=0}^{\infty} T^{-i}\mathcal{B}$. This implies, by (31.2), that h(T) = 0.

Consider now the measure space with the points 0,..., k-1 (k = 1, 2, 3,...) bearing the measures $p_0,...,p_{k-1}$. Let X be the bilateral sequence space based on that space, let P be the usual product measure in X , and let T be the bilateral shift on X . If $\mathcal{B}$ is the field generated by the sets $\{x: x_0 = i\}$, i = 0,..,k-1, then $\bigvee_{i=-\infty}^{+\infty} T^i\mathcal{B} = \mathcal{S}$ and therefore, by (31.1), h(T) = h($\mathcal{B}$, T). Since the fields $T^{-i}\mathcal{B}$, i = 0,..., n-1 are independent, it follows that

$$\bar{H}(\bigvee_{i=0}^{n-1} T^{-i}\mathcal{B}) = n\bar{H}(\mathcal{B}) ,$$

and hence that

$$h(\mathcal{B}, T) = \bar{H}(\mathcal{B}) .$$

This implies, finally, that

$$h(T) = -\sum_{i=0}^{k-1} p_i \log p_i .$$

If, in particular, $p_0 = ... = p_{k-1} = \frac{1}{k}$, then h(T) = log k . This proves that the 2-shift is not conjugate to the 3-shift.

CENTRO INTERNAZIONALE MATEMATICO ESTIVO
(C.I.M.E.)

E B E R H A R D   H O P F

SOME TOPICS OF ERGODIC THEORY

ROMA - Istituto Matematico dell'Università - 1960

SOME TOPICS OF ERGODIC THEORY

by

E. HOPF [1)]

INTRODUCTION. The first topic discussed in these lectures is a general ergodic theorem for positive linear transformations of certain general function spaces into themselves. The author conjectured this theorem sometime ago but it was proved only recently by Chacon and Ornstein. It contains as special cases the wellknown individual ergodic theorems in the case when the linear function-transformation is induced by a point-transformation of the underlying point-space onto itself. The theorem is presented in two equivalent forms (first and second ergodic theorem) the first of which deals mainly with additive set functions whereas the second expresses it in terms of point functions. There is no denying that the second form of the theorem has at present greater immediate appeal than the first. However, the theorem in that second form actually deals with equivalence classes of point-functions (i.e. with set functions) rather than point-functions themselves, and for this reason the first form of the theorem has been included here. We also consider implications of the theorem. We discuss the case where the operator is induced by a point-transformation and we consider here not only one-to-one (classical case) but also many-to-one mappings. The latter case is interesting because the ergodic theorem deals with operators in the first place and because, in general, a many-to-one mapping may be

──────────────────────────

1)
. The author expresses his appreciation to his Italian colleagues
for their invitation and to the Office of Naval Research, Washing-
ton, D.C., for their support of this work.

49

E.Hopf

regarded as an operator in many different ways. Operators corresponding to the same mapping may have widely differing spectral characteristics.

A lot of ingenuity has been applied to the generalization of the classical ergodic theorems and less attention has been given to the ergodic behaviour in specific cases, specific mappings and specific differential systems. There are various open problems which are undoubtedly of considerable mathematical interest. It is true that the solutions of the classical problem of three bodies behave "in general" (apart from a set of measure zero in phase space) such that at least one of the bodies goes infinitely far away from the center of gravity as $t \to \infty$?

In the language of ergodic theory, is the general problem of three bodies of dissipative type? In certain limit cases of the problem this can certainly not be so, for instance, in Hill's lunar theory where one mass is zero (moon), another (sun) infinitely large and infinitely far away from the earth which is the origin of a specially chosen rotating coordinate system. For certain values of the moon's energy the manifold of constant energy in phase space consists of two disjoint parts one of which has finite phase volume. In this part the phase motion cannot possibly be of dissipative type. It is metrically transitive in this part or, in other words, does the moon "in general" run through all positions of the part with mean time of sojourn proportional to the volume of the portion? While such questions can hardly claim astronomical interest (the time intervals for predictions of ergodic theory to come true are probably so large that the Newtonian three body system is no longer a good idealization) they are, mathematically, justified and interesting. The reason for that is that the actual stabi-

E.Hopf

lity of the classical periodic motions is doubtful in spite of their infinitesimal stability. Such questions are not new. They have been voiced already by K.Schwarzschild and G.D.Birkhoff. They are also very difficult, and we have no intention to discuss them in detail in these lectures. However, there is another circle of differential problems in which the ergodic properties of the solutions are comparatively easy to determine, namely, the geodetic flow on complete surfaces of negative curvature. These surfaces furnish many interesting examples for ergodic theory. We present this subject here essentially in the same simple way in which we had treated it in our original memoir [11]. The main idea there has been condensed here into a "principal lemma".

1a. THE ERGODIC THEOREM FOR POSITIVE LINEAR OPERATORS.

Consider a fixed space X with points x, y,... and a fixed σ-field $\mathcal{F}$ of subsets A, B,... of X . X itself is supposed to belong to $\mathcal{F}$, $X \in \mathcal{F}$. Real-valued point functions in X are denoted by small Latin letters, $f = f(x)$, $g = g(x),\ldots$. All point functions considered in this paragraph are $\mathcal{F}$-measurable, either by assumption or by implication. Real-valued set functions in $\mathcal{F}$ which are finite and σ-additive are denoted by small Greek letters, $\phi = \phi(A)$, $\psi = \psi(A),\ldots$. A finite σ-additive set function π in $\mathcal{F}$ which is ≥ 0 is a finite measure in $\mathcal{F}$.

By $\emptyset$ we denote the totality of the finite σ-additive set functions ϕ defined in $\mathcal{F}$. $\emptyset$ forms a Banach space with the norm

$$\|\phi\| = \text{total variation of } \phi \text{ in } X \ .$$

We use the symbol

$$(1) \qquad \phi \prec \pi \ , \ \pi \geq 0 \ ,$$

E. Hopf

in the sense that $\pi(A) = 0$, $A \in \mathfrak{F}$, $\Rightarrow$ $\phi(A) = 0$ (ϕ is absolutely continuous with respect to the measure π). Throughout this paper the left hand side is $\in \emptyset$. The right is always a measure (defined in $\mathfrak{F}$) but not necessarily finite (not necessarily $\in \emptyset$). However we require π to be σ-finite: X is the union of countably many sets ($\in \mathfrak{F}$) with $\pi < \infty$. We are mainly interested in finite σ-additive set functions because they arise naturally from the applications to Markov processes (mentioned farther below) but other applications (to infinite invariant measures of point-transformations) require the occasional use of possibly infinite but σ-finite measures μ. For a given σ-finite measure μ defined in $\mathfrak{F}$ we make use of the notation

$$(2) \qquad \emptyset_\mu = [\phi \,|\, \phi \in \emptyset , \ \phi \prec \mu] \ .$$

Consider now a linear operator T which acts on members of $\emptyset$ and which produces again members of $\emptyset$, $\phi' = T\phi$. In view of the most important applications it would not be natural to demand of T that it be defined in all of $\emptyset$, but of the subspace $\emptyset'$ in which it is defined we require that it has the following closure properties (see [14], § 2) :

$\emptyset_1$) $\emptyset'$ is a linear space. $\emptyset_2$) $\emptyset'$ is closed under contractions $\phi \rightarrow \phi_B$, $\phi_B(A) = \phi(A-B)$, $B \in \mathfrak{F}$. $\emptyset_3$) $\phi_n \in \emptyset'$, $\phi_n \prec \phi$, ϕ finite in $\mathfrak{F}$, implies that $\phi \in \emptyset'$.

$\emptyset$ itself has all these properties and so does every space $\emptyset_\mu$ where μ is a σ-finite measure in $\mathfrak{F}$. The latter fact reflects well known facts of the theory of integration if it is translated by means of the isometry between $\emptyset_\mu$ and L_μ stated farther below. It is also true that

$$\pi \in \emptyset' , \quad \pi \geq 0 \Rightarrow \emptyset_\pi \subset \emptyset' \ .$$

(see [14], § 3). Of the operator T we demand that

E.Hopf

T_1) $T\emptyset' \subset \emptyset'$. T_2) T is linear. T_3)T is positive :
$\varphi \geq 0 \Rightarrow T\varphi \geq 0$. T_4) $\|T\phi\| \leq \|\phi\|$, $\phi \in \emptyset'$.

The last condition means simply that $\|T\| \leq 1$. In order that a positive linear operator T satisfy this condition it is necessary and sufficient that ($T\phi(A)$ denotes the value of $T\phi$ at A)

$$T'_4) \quad \phi \geq 0 \Rightarrow T\phi(X) \leq \phi(X) .$$

The necessity is trivial as $\|\phi\| = \phi(X)$, $\phi \geq 0$. That the condition is sufficient follows if we use the well-known fact ([7], Chap.VI) that every $\phi \in \emptyset'$ is the difference of two finite measures $\phi = \phi^+ - \phi^-$ (ϕ^+ and $-\phi^-$ are contractions of ϕ and therefore $\in \emptyset'$) such that $\|\phi\| = \phi^+(X) + \phi^-(X)$. In fact

$$T\phi = T\phi^+ - T\phi^-, \quad \|T\phi\| \leq \|T\phi^+\| + \|T\phi^-\|$$

and, by virtue of the positivity of T and of T'_4),

$$\|T\phi\| \leq T\phi^+(X) + T\phi^-(X) \leq \phi^+(X) + \phi^-(X) = \|\phi\| .$$

The most important operators of this kind are the Markov operators, the positive linear operators T with the property that

$$T\phi(X) \equiv \phi(X) .$$

They are intimately associated with Markov processes in X : If $\phi \geq 0$, $\phi(X) = 1$, is a probability distribution in X then $T\phi$ is the probability distribution which results through the process. A Markov operator possesses a representation

$$(3) \qquad T\phi(A) = \int_X P(x,A)\phi(dx)$$

where $P(x,A)$ can be interpreted as the probability of transition of x into A ([14], §2). The proof given in [14] works also for

53

E.Hopf

the general T and furnishes the same representation (3) where P
is still $\geq$ 0 and σ-additive in A but where

$$P(\chi, X) \geq 1$$

instead of being $\equiv$ 1.

We want to state two equivalent ergodic theorems, the first
in terms of set functions and the second in terms of point functions. The proof of the second theorem is given in the following
paragraph. Here we discuss solely their meaning and we prove their
equivalence. In order to state the first we need a weak form of
the splitting of a finite σ-additive set function ϕ into absolutely continuous and singular part with respect to a finite or σ-fi-
nite measure : There exists a (perhaps vacuous) set N such that

$$\pi(N) = 0 \ , \quad \phi \prec \pi \quad \text{within X-N} \ .$$

Within X-N there exists therefore a Radon-Nikodym derivative of
ϕ with respect to π. It is, of course, determined only up to a
set with π = 0. The ambiguity in the choice of N and, also, of the
derivative in X-N will have no effect in what follows.

Let us now return to the situation faced above. We have a
linear space $\emptyset'$ of finite σ-additive set functions as stipulated
and a positive linear operator T of $\emptyset'$ into $\emptyset'$. Let $\phi \in \emptyset'$, $\pi \in \emptyset'$,
$\pi \geq$ 0. The ergodic theorem is concerned with the sequence of pairs
of set functions (the second is $\geq$ 0 as T is positive)

$$(4) \qquad S_n\phi = \sum_{o}^{n-1} T^\nu \phi \ , \ S_n\pi = \sum_{o}^{n-1} T^\nu \pi \geq 0 \ ,$$

n = 1, 2,... . They are, of course, all in $\emptyset'$. Take, for each n,
any set N_n such that

E.Hopf

(5) $\qquad S_n \pi(N_n) = 0 \ , \ S_n \phi \prec S_n \pi \ $ within $ \ X-N_n$

and take any Radon-Nikodym derivative $q_n(x)$ (within $X-N_n$) of the first with respect to the second set function (4),

(6) $\qquad S_n \phi(A) \equiv \int_A q_n(x) \, S_n \, \pi \, (dx), \ A \subset X-N_n \ .$

(5), first part, means that

(7) $\qquad T^\nu \pi(N_n) = 0, \quad \nu = 0,1,\ldots, n-1 \ .$

The ergodic theorem is concerned with the pointwise limit behaviour of the sequence of Radon-Nikodym derivatives

$$q_1(x), \ q_2(x), \ \ldots \ .$$

The only points x that matter in this context are those in which at least a tail end of this sequence is well defined. They are precisely the points of the set

$$\bigcup_1^\infty \bigcap_n^\infty (X-N_i) = \bigcup_1^\infty (X - \bigcup_n^\infty N_i) = X - \bigcap_1^\infty \bigcup_n^\infty N_i$$

or, in other words, of the set

(8) $\qquad X-N', \quad N' = \bigcap_1^\infty N'_n \ , \quad N'_n = \bigcup_n^\infty N_i \ ,$

(7) implies that

$$T^\nu \pi (N'_n) = 0 \ , \quad 0 \leq \nu < n \ ,$$

and finally, as $N' \subset N'_n$, $n \geq 1$, that

(9) $\qquad T^\nu \pi (N') = 0, \quad \nu = 0, \ 1, \ldots \ .$

Result : The Radon-Nikodym derivatives $q_n(x)$, $n = 1, \ 2, \ldots$, are at least tailwise defined outside some set N' which satisfies (9).

55

E.Hopf

The same argument makes it clear that they are not only defined but completely determined in this sense: Any other choice of sets N_n and derivatives q_n affects values in tails only in a set with the property (9). In particular, the limit behaviour of the $q_n(x)$, $n \to \infty$, is independent of such a choice outside of some such set. With these facts in mind we can now state the

FIRST ERGODIC THEOREM. Let $\emptyset'$ denote a linear space of finite σ-add. set functions as stipulated above and let T be a positive linear operator from $\emptyset'$ to $\emptyset'$. Let $\phi \in \emptyset'$, $\pi \in \emptyset'$, $\pi \geq 0$. Consider the operators $S_n = \sum_{0}^{n-1} T^{\nu}$ and denote, for each $n = 1, 2, \ldots$, by N_n any set with the property (5) and by $q_n(x)$ any Radon-Nikodym derivative in $X - N_n$ of $S_n \phi$ with respect to $S_n \pi$. Then, if T has a norm $\| T \| \leq 1$,

$$\lim_{n \to \infty} q_n(x)$$

exists and is finite in each point outside some set N which contains the set N' defined by (8) and which satisfies $T^{\nu} \pi(N) = 0$, $\nu = 0, 1, \ldots$.

In virtue of what was said before it is clear that in order to prove the theorem, it suffices to prove it for *some* sequence, N_n, q_n. We have to remember this later.

An important case of this theorem is the one in which $\phi \prec \pi$ (in X): It was shown in [14], § 3, that (both in X)

$$(10) \qquad \phi \prec \pi \implies T\phi \prec T\pi \ .$$

For the sake of completeness the simple proof of (10) is mentioned here a little farther below. From (10) it follows that

$$\sum_{0}^{n-1} T^{\nu} \phi \prec \sum_{0}^{n-1} T^{\nu} \pi \quad \text{in} \quad X \ .$$

E.Hopf

In fact, the vanishing of the right hand side for some set A implies the vanishing of each of its terms and, by virtue of (10), $T\pi(A) = 0 \Rightarrow T^{\nu}\phi(A) = 0$ which in turn implies that the left hand side is zero. So, under the additional hypothesis $\phi \prec \pi$, all the derivatives $q_n(x)$ may be taken as defined in all of X and $N' = 0$ may be assumed.

We add here the proof of (10). Use the representations (3) of $T\phi$ and $T\pi$ and observe that

$$T\pi(A) = 0 \Rightarrow \pi(B) = 0, \quad B = [x \mid P(x,A) > 0] \ .$$

Therefore, $\pi(B') = 0$, $B' \subset B$. From the premise in (10) it follows now that $\phi(B') = 0$, $B' \subset B$. This and the representation of T make it obvious that $T\phi(A) = 0$. (10) is hereby proved.

It was mentioned above that the simplest of the spaces $\emptyset'$ considered is a space $\emptyset_{\mu}$ where μ is a finite measure, simplest in the sense that $\mu \in \emptyset' \Rightarrow \emptyset_{\mu} \subset \emptyset$ and that every $\emptyset_{\mu}$ is itself a space $\emptyset'$. The following remark is technically useful to us.

LEMMA. Under the hypotheses of the first theorem there exists a subspace $\emptyset_{\mu}$ of $\emptyset'$ with some finite measure $\mu \in \underline{\emptyset}'$ such that the relations mentioned in these hypotheses stay true with $\emptyset_{\mu}$ in place of $\emptyset$.

To see this we must prove the existence of some $\mu \in \emptyset'$ such that

$$\tag{11} \phi \in \emptyset_{\mu} \ , \quad \pi \in \emptyset_{\mu} \ , \quad T\emptyset_{\mu} \subset \emptyset_{\mu}$$

where ϕ and π are the set functions given in the hypothesis of the theorem. For this purpose we use again the representation $\phi = \phi^{+} - \phi^{-}$ of ϕ as a difference of two finite measures which are both in $\emptyset'$. We claim that the measure defined by

57

E.Hopf

$$\mu = \sum_1^\infty 2^{-\nu} \, T^\nu \, \pi' \; , \; \pi' = \phi^+ + \phi^- + \pi \; ,$$

fits the bill. μ is finite because (see T_4')) $T^\mu \pi'(X) \leq \pi'(X)$.

Obviously, $\phi \prec \pi'$, $\pi \prec \pi'$ and

$$(12) \qquad \phi \prec \mu \; , \quad \pi \prec \mu \; , \quad T\mu \prec \mu \; .$$

The first two relations are, by virtue of (2), identical with the
first two relations (11), and the third of the latter relations
follows from the third relation (12) and from the basic fact (10)
as applied to ϕ, μ.

In order to formulate the ergodic theorem in terms of point
functions we start from a fixed measure μ (defined in $\mathfrak{F}$) which
need not be finite but which is supposed to be σ-finite. Consider
the Banach space L_μ with the norm

$$\| f \| = \int_X |f(x)| \, \mu(dx)$$

and denote by T a positive linear operator from L_μ to L_μ with a
norm $\|T\| \leq 1$. Strictly speaking, argument and value of T are
both μ-equivalence classes of μ-integrable functions $f(x)$ in X.
We may, however, regard f and f' in $f' = Tf$ as functions if we
keep in mind that f' is determined only up to a set with $\mu = 0$
and that a change of f in a set with $\mu = 0$ has no effect on the
μ-equivalence class f'.

SECOND ERGODIC THEOREM. Let μ be a σ-finite measure in X ,
$\mathfrak{F}$ and let T be a positive linear operator from L_μ to L_μ with
a norm $\|T\| \leq 1$. Suppose that $f \in L$, $p \in L$, $p \geq 0$. Then the quo-
tients

$$q_n(x) = q_n(x; \, f, \, p) = \frac{\sum_0^{n-1} T^\nu f(x)}{\sum_0^{n-1} T^\nu \phi(x)}$$

E.Hopf

n = 1, 2,... , have a finite limit in each point outside some set N which contains the set N' where $\sum_{0}^{\infty} T^{\nu} p(x) = 0$ and which satisfies $\mu(N-N') = 0$.

Before discussing the implications of these theorems we show that *the first theorem for the case $\emptyset' = \emptyset_\mu$ where μ is a given σ-finite measure in $\mathcal{F}$ and the second theorem are merely different forms of the same fact.* This implies the equivalence in question because, by virtue of the lemma, the full first theorem is already a consequence of its special case in which $\emptyset' = \emptyset_\mu$ with μ being a finite measure in $\mathcal{F}$.

The proof of equivalence rests upon the one-to-one linear isometry between a space $\emptyset_\mu$ with a given σ-finite measure μ and the space L_μ of all μ-integrable point functions in X or, rather, of the μ-equivalence classes of these functions. The isometry is furnished by the relations

$$(13) \qquad \phi(A) = \int_A f(x)\ \mu(dx)\ , \qquad A \in \mathcal{F}\ ,$$

(f = Radon-Nikodym derivative of ϕ w.r. to μ) and

$$\|\phi\| = \int_A |f(x)|\ \mu(dx) = \|f\|\ .$$

A linear operator T from L_μ to L_μ induces by means of the formula

$$(14) \qquad T\phi(A) = \int_A Tf(x)\ \mu(dx)$$

a linear operator T from $\emptyset_\mu$ to $\emptyset_\mu$ and vice versa. The use of the same letter T in both spaces is permitted since the distinction in the notation for set functions and point functions makes it clear in each case which T is meant. It is also clear that positivity of T in one space implies the positivity in the other. Since corresponding functions ϕ, f have the same norm the two T's have the

E.Hopf

same norm. For either T we use the positive linear operators

$$(15) \qquad S_n = \sum_{o}^{n-1} T^{\nu} , \qquad n > 0 ,$$

which had already been introduced before in one case. For corresponding functions ϕ , f there holds

$$(16) \qquad S_n \, \phi(A) = \int_A S_n f(x) \, \mu(dx) , \qquad A \in \mathfrak{F} ,$$

and, for corresponding non-negative functions π, p,

$$(17) \qquad S_n \, \pi(A) = \int_A S_n p(x) \, \mu(dx) , \qquad A \in \mathfrak{F} .$$

If, for given π, p, a set N_n is defined by

$$(18) \qquad N_n = [x \,|\, S_n p(x) = 0] = [x \,|\, T^{\nu} p(x) = 0; \ \nu = 0,\ldots,n-1]$$

it is obvious that $\mu \prec S_n \pi$ holds within $X-N_n$. Consequently,

$$(19) \qquad S_n \pi(N_n) = 0 , \qquad S_n \, \phi \prec S_n \, \pi \quad \text{within } X-N_n$$

If we define

$$(20) \qquad q_n(x) = \frac{S_n f(x)}{S_n p(x)}$$

in $X-N_n$ then we can infer from (16) and (17) that

$$S_n \phi(A) = \int_A q_n(x) \, S_n p(x) \, \mu(dx) = \int_A q_n(x) \, S_n \, \pi \, (dx)$$

holds for any $A \in \mathfrak{F}$, $A \subset X-N_n$. In other words, $q_n(x)$ as defined by (20) is a Radon-Nikodym derivative of $S_n \phi$ with respect to $S_n \pi$ within $X-N_n$. The sets (18) form a descending sequence Therefore, the sets N_n', N' formed by them through (8) are $N_n' = N_n$ and

$$(21) \qquad N' = [x \,|\, \sum_{o}^{\infty} T^{\nu} p(x) = 0] .$$

To this we add the remark that

E.Hopf

$$(22) \qquad N \supset N' , \quad \sum_{o}^{\infty} T^{\nu} \pi(N) = 0 \iff N \supset N', \quad \mu(N-N') = 0$$

where N' is the set (21) and where $\pi \geq 0$ and $p \geq 0$ are corresponding functions,

$$\pi(A) = \int_A p(x) \, \mu(dx) , \qquad A \in \mathcal{F} .$$

The truth of this is evident from (21) and from the identity

$$\sum_{o}^{\infty} T^{\nu} \pi(N) = \int_{N'} \sum_{o}^{\infty} T^{\nu} p(x) \, \mu(dx) + \int_{N-N'} \sum_{o}^{\infty} T^{\nu} p(x) \, \mu(dx).$$

Now it is obvious that the second theorem and the special
first theorem, $\emptyset' = \emptyset_{\mu}$ where μ is the μ in the premise of the second theorem, say exactly the same thing in the following sense.
The functions f, ϕ and p, π as well as the two operators T correspond to each other, respectively, under the isometry between L_{μ}
and $\emptyset_{\mu}$, and the first theorem is understood to be stated for some
special sequence of sets N_n satifsying (5) and Radon-Nikodym derivatives $q_n(x)$ in $X-N_n$, namely for the N_n defined by (18) and the
q_n defined by (20). Remember here that the first theorem necessarily holds for any sequence N_n, q_n satisfying (5) if it holds for some such sequence. The equivalence is hereby established.

1b. IMPLICATIONS OF THE ERGODIC THEOREM.

The case $\|T\| < 1$ is of no interest because then every series

$$S_{\infty} f = \sum_{1}^{\infty} T^k f$$

itself converges in norm and, therefore, almost everywhere. This
situation can arise even in the limit case $\|T\| = 1$ and even in the
case of a Markov operator T. Closer attention will be paid to the
question of convergence or divergence farther below. Let us briefly examine how much the theorem says for general Markov operators T.

E.Hopf

There are two extreme cases of Markov processes. In the first case the transition from one stage to the next possesses a smooth spread. More precisely,

$$P(x, \ A) = \int_A K(x, \ y) \ \mu(dy)$$

where K is a continuous function of both arguments. In this case, considerably sharper theorems exist about the limit behaviour of the powers T^n. The same is true if, more generally, the operator T is completely continuous in some topology naturally connected with the process (the uniform ergodic theorem of Yoshida and Kakutani [20]). At the other extreme of Markov processes there is the strictly deterministic case, sharp point to point transition. It is in this case and in cases closely bordering on it that the theorem has real significance.

Let T denote a single-valued point transformation $x' = Tx$ of the space X precisely into X ,

$$TX = X .$$

Again, the use of the same letter T in various contexts will not cause confusion if close attention is paid to the argument of T . Tx need not be one-to-one. It is clear what is meant by TA, $A \subset X$. We define $T^{-1}A$ by $[x| \ Tx \in A]$. Observe that T^{-1} exactly preserves the elementary set operations, formation of union and intersection, and also the property of disjointness of sets. If T is one-to-one the same is true for T itself whereas this is in general not the case if T^{-1} is many-valued. We wish to associate a Markov operator with each of the point transformations T and T^{-1}.

Consider first T^{-1} (regarded as a transformation of sets) and suppose that T^{-1} is $\mathfrak{F}$-measurable, in other words that it sends sets of the σ-field $\mathfrak{F}$ again into sets of $\mathfrak{F}$. We may express this by writing

E.Hopf

$$(23) \qquad T^{-1}\mathcal{F} \subset \mathcal{F} \quad .$$

If we express the desired Markov operator as an operator on finite σ-additive set functions (i.e. on elements ϕ of $\emptyset$) then the answer is simply

$$(24) \qquad T'\phi(A) = \phi(T^{-1}A) \quad , \quad \phi \in \emptyset \ , \quad A \in \mathcal{F} \quad .$$

In fact, linearity and positivity are obvious, and the right hand side is an element of $\emptyset$ whenever ϕ is. We may, therefore, apply the first ergodic theorem with $\emptyset' = \emptyset$, in particular, the case where $\phi \prec \pi$. The result is essentially the ergodic theorem of Hurewicz [16]:

Let $\mathcal{F}$ be a σ-field of subsets of X including X itself. Suppose that T is a single-valued transformation of X into precisely X such that T^{-1} is $\mathcal{F}$-measurable. Suppose, furthermore, that ϕ, π are finite σ-additive set functions in $\mathcal{F}$ and that $\pi \geq 0$, $\phi \prec \pi$. Then the Radon-Nikodym derivative $q_n(x)$ of $\sum_{o}^{n-1} \phi(T^{-\nu}A)$ with respect to $\sum_{o}^{n-1} \pi(T^{-\nu}A)$ has a finite limit as $n \to \infty$ in every point x of X-N where $\sum_{o}^{\infty} \pi(T^{-\nu}N) = 0$.

Of course, if T is one-to-one and $\mathcal{F}$-measurable then the theorem holds with all the negative powers of T replaced by the corresponding positive powers. This was the case considered by Hurewicz. If, however, T^{-1} is not single-valued then it is more interesting to consider T itself.

To find a Markov operator associated with T itself is less easy. $T\phi(A) = \phi(TA)$ cannot be the answer because the right hand side need not be additive in A. The answer which is this time best expressed in terms of point functions is obtained in the following way. We assume again (23) : T^{-1} sends sets of $\mathcal{F}$ again into sets

E.Hopf

of $\mathfrak{F}$. We assume furthermore, the existence of a σ-finite measure μ in $\mathfrak{F}$ such that $\mu(A)$ is absolutely continuous with respect to the measure $\mu(T^{-1}A)$ and vice versa. If we observe that $T^{-1}\mathfrak{F}$ is again a σ-field and that $\mu(TA')$ is a measure in $T^{-1}\mathfrak{F}$ (but not in $\mathfrak{F}$) we can evidently phrase that assumption as follows. $\mu(TA')$ is absolutely continuous with respect to $\mu(A')$ within $T^{-1}\mathfrak{F}$ and vice versa. The Radon-Nikodym theorem therefore implies the existence of a $T^{-1}\mathfrak{F}$ -measurable and almost everywhere finite function $r(x) \geq 0$ such that

$$(25) \qquad \mu(TA') = \int_{A'} r(x)\mu(dx) , \qquad A' \in T^{-1}\mathfrak{F} .$$

The assumption of the absolute continuity of $\mu(T^{-1}A)$ with respect to $\mu(A)$ implies that $r > 0$. The condition of $T^{-1}\mathfrak{F}$ -measurability determines $r(x)$ uniquely in the sense of μ-equivalence. However, if this condition is weakened to the condition of $\mathfrak{F}$ -measurability - the latter is weaker by virtue of (23) - then there exist also other functions $r(x) \geq 0$ which satisfy (25). An example illustrates this farther below. What does $T^{-1}\mathfrak{F}$ -measurability mean for r? It means that every set $r \geq c$ is of the form $T^{-1}A$. This in turn means that $r(x)$ is of the form $s(Tx)$ where $s(y)$ is single-valued and $\mathfrak{F}$ -measurable in X. (25) may be written

$$(25') \qquad \mu(A) = \int_{T^{-1}A} r(x) \mu(dx) = \int_{X} r(x) \, e_{A}(Tx)\mu(dx), \quad A \in \mathfrak{F} ,$$

where e_{A} is the characteristic function of A.
In other words, the relation

$$(26) \qquad \int_{X} f(x) \, \mu(dx) = \int_{X} r(x) \, f(Tx) \, \mu(dx)$$

holds for every $\mathfrak{F}$ -measurable characteristic function. Well-known arguments show then that (26) must hold for every $\mathfrak{F}$ -measurable

E.Hopf

and μ-integrable function $f(x)$. Simultaneously the integrability of the right hand integrand is inferred. We can express these results simply by saying that the positive linear operator

$$(27) \qquad Tf(x) = r(x) \, f(Tx)$$

is a Markov operator from L_μ to L_μ as it was defined in connection with the second theorem. To complete the story it must be shown that (27) is actually an operator between μ-equivalence classes, that it transforms a function that vanishes except in a set with $\mu = 0$ again into such a function. This, however, follows at once from the assumption that $\mu(T^{-1}A)$ is absolutely continuous with respect to $\mu(A)$. The second ergodic theorem can now be applied.

Let T denote s single-valued mapping of X onto X and let $\mathfrak{F}$ be a σ-field of subsets of X such that $A \in \mathfrak{F} \Rightarrow T^{-1}A \in \mathfrak{F}$. Let, furthermore, μ be a measure on $\mathfrak{F}$ such that $\mu(A)$ and $\mu(T^{-1}A)$ are both σ-finite and absolutely continuous with respect for each other. Determine an $\mathfrak{F}$-measurable, finite function $r(x) \geq 0$ such that (26) is satisfied by every $\mathfrak{F}$-measurable $f(x) \in L_\mu$ and let

$$r_0(x) = 1 \, , \quad r_n(x) = r(T^{n-1}x)r_{n-1}(x) \, , \quad n > 0 \, .$$

Then, if $f \in L_\mu$, $p \in L_\mu$, $p > 0$ the quotients

$$q_n(x) = \frac{S_n f(x)}{S_n p(x)} \quad , \qquad S_n g(x) = \sum_0^{n-1} r_i(x) \, g(T^i x)$$

have a finite limit as $n \to \infty$ in each point of $X-N$ where $\mu(N) = 0$.

The hypothesis that $p > 0$ can, of course, be relaxed to : $S_\infty p(x) > 0$. If T is one-to-one this is the ergodic theorem of Halmos [6] or, rather, its generalization by Oxtoby [19]. Halmos supposed that $S_\infty p = \infty$. We mention, by the way, that the theorem as stated above was established by us in 1946 in an unpublished paper.

65

E.Hopf

In the important case in which the measure μ is invariant under T (or, rather, under T^{-1}),

$$(28) \qquad \mu(T^{-1}A) = \mu(A), \quad A \in \mathfrak{F} \, ,$$

we may take r = 1 and

$$S_n g(x) = \sum_0^{n-1} g(T^i x)$$

In this case the theorem is the author's generalization of G.D. Birkhoff's ergodic theorem, and in the case where the invariant measure is finite, $\mu(x) < \infty$, it becomes the latter theorem itself because then $p = 1 \in L_\mu$. In these theorems, however, only one-to-one transformations T were considered.

Consider the following example of a two-to-one transformation (Caratheodory [1]). X is the circular line of all numbers x mod 2π, $\mathfrak{F}$ is the field of all Borel sets in X . T is defined by $x' = 2x$ mod 2π . $T^{-1}\mathfrak{F}$, the totality of all sets $T^{-1}A$, $A \in \mathfrak{F}$, consists of all sets A which are invariant under the covering transformation $x^* = x + \pi$ mod 2π. The ordinary Borel measure μ in X is invariant under T^{-1}. (25) says that

$$r(x) + r(x + \pi) = 2$$

holds for almost all x . The solution r = 1 is essentially the only one which is $T^{-1}\mathfrak{F}$ -measurable, in other words, which is measurable and invariant under the covering transformation, $r(x + \pi) = r(x)$. There are, however, infinitely many $\mathfrak{F}$ -measurable solutions so, for example, $r = 2 \sin^2(x/2)$. The ergodic theorem says that

$$\frac{1}{n} \sum_0^{n-1} f(2^\nu x)$$

E.Hopf

has a finite limit almost everywhere if f is of period 2π and integrable.

Let us return, for a moment, to the general second theorem with T being a Markov operator between point functions $\in L_\mu$ and with μ being a given measure. What does it mean to say that μ is an invariant finite measure of the Markov process? As the associated operator T on set functions ϕ is determined by

$$\phi(A) = \int_A d\mu \ , \quad T\phi(A) = \int_A Tf d\mu$$

it means that $T\mu = \mu$ or, in other words, that the function $f = 1$ is $\in L_\mu$ (finiteness) and that $T1 = 1$ (invariance). In this case the second theorem states the convergence almost everywhere of

$$\frac{1}{n} \sum_0^{n-1} T^\nu f(x) \ , \quad f \in L_\mu \ .$$

L_μ is the largest function space in which this theorem can be expected to hold. So, for instance, if $Tf(x) = f(Tx)$ with T one-to-one and μ-preserving and T metrically transitive the limit in question is $\int_X f d\mu / \mu(X)$ which limit would be ∞ if $f \geqslant 0$ and $\int_X f d\mu = \infty$. The theorem was proved by us for the full space L_μ [14]. Previously, it had been proved for smaller spaces, by Kakutani [17] for bounded f and by Doob [3] for $f \in L^p$, $p > 1$. In our paper [14] we had stated the general second ergodic theorem for Markov operators as a conjecture. It was, however, only very recently that the second theorem was proved by Chacon and Ornstein [2].

10. THE WORK OF CHACON AND ORNSTEIN. PROOF OF THE SECOND ERGODIC THEOREM.

What follows is a simplified presentation [15] of the ingeneous proof of these authors.

Since $X, \mathcal{F}$, μ are given once for all we need not refer to them explicitely. We write $\int_A f$ and simply $\int f$ if $A = X$. The proof makes use of the lattice operations $f \to f^+$, $f \to f^-$ which carry $L = L_\mu$ into itself,

$$(29) \qquad f^+(x) = \sup(f(x),0) \; ; \; f^-(x) = -\inf(f(x),0).$$

There holds

$$(30) \qquad f = f^+ - f^- \; , \qquad |f| = f^+ + f^- \; .$$

f^+, f^- are both ≥ 0 but never both > 0.

$$(31) \qquad f = f_2 - f_1, \; f_1 \geq 0 \Rightarrow f_2 - f^+ \geq 0 \; , \; f_1 - f^- \geq 0$$

follows from this remark as at least one of the two last inequalities must hold and as their left hand sides are equal. From $Tf = Tf^+ - Tf^-$, the positivity of T and from (31) we infer that

$$(32) \qquad (Tf)^+ \leq Tf^+ \; , \; (Tf)^- \leq Tf^- \; , \; |Tf| \leq T|f| \; .$$

Recall that $\|T\| \leq 1$ is equivalent to the statement that

$$(33) \qquad g \geq 0 \Rightarrow \int Tg \leq \int g$$

The proof of the second ergodic theorem is carried through in several steps. The first (principal) lemma we formulate as follows (statement and proof are somewhat more transparent than in Chacon's and Ornstein's paper).

E.Hopf

LEMMA 1. If $f \in L$ and if

$$\sup_{n>0} S_n f(x) > 0 \quad,$$

$S_n = \sum_0^{n-1} T^i$, holds in each point x of a set $A(\in \mathcal{F})$ then, to e-very $\xi > 0$, there exist functions $h \in L$, $g \in L$ such that

$\alpha)\quad h^- \le f^-\ ,\qquad\qquad \beta)\quad h = f + Tg - g\ ,\quad g \ge 0\ ,$

$\gamma)\quad \int h \le \int f\ ,\qquad\qquad \delta)\quad \int_A h^- < \epsilon\ .$

PROOF. The role of $\beta)$ will emerge farther below: The additional term $Tg-g$ has no influence on the limit behaviour of $q_n = S_n f/S_n p$. $\delta)$ means that h is practically ≥ 0 in A. $\alpha)$ implies that $h \ge f$ as long as $f < 0$ whereas $\gamma)$ implies a limitation on h in the opposite sense.

To prove the lemma we define a sequence of functions $h_i \in L$ recursively, by applying T to the positive part only,

$$(34)\qquad h_{i+1} = Th_i^+ - h_i\ ,\quad h_0 = f\ ,$$

and we show that each h_i has the properties $\alpha) - \gamma)$ and that it satisfies $\delta)$ for sufficiently large i. From (34), the positivity of T and from (31) it follows that

$$(35)\qquad h_{i+1}^- \le h_i^-\ ,$$

$i \ge 0$, which implies $\alpha)$ as $h_0 = f$. (34) may be written

$$h_{i+1} = h_i + Th_i^+ - h_i^+$$

Hence, by summation,

$$(36)\qquad h_{i+1} = f + Tg_i - g_i\ ,\quad g_i = \sum_0^i h_\nu^+$$

E.Hopf

and this proves β). β) implies γ) by virtue of (33).
If (36) is combined with

$$g_{i+1} - g_i = h^+_{i+1} \geq h_{i+1}$$

there follows

$$g_{i+1} \geq f + Tg_i \; ,$$

$i \geq 0$. As T is order-preserving this implies, by induction, that $(n > 0)$

$$g_n \geq \sum_o^{n-1} T^i f + T^n g_o$$

and, since $g_o = h^+_o = f^+ \geq f$, that

$$\sum_o^n h^+_i = g_n \geq \sum_o^n T^i f \; .$$

Obviously, this holds for $n = 0$ too. This inequality and the hypothesis of the lemma imply now : In each point of A at least one of the h^+_i, $i \geq 0$, is > 0, or, in other words, at least one of the h^-_i is = 0. Hence and from $h^-_i \geq 0$ and from (35) it follows that $h^-_n \downarrow 0$ holds in A and, consequently, that $\int_A h^-_n \to 0$. Lemma 1 is hereby completely proved.

It is possible to eliminate the auxiliary function h of Lemma 1 by means of the properties α) - δ). The result is the
LEMMA 2. If $f \in L$ and if

$$\sup_{n>0} S_n f(x) \geq 0 \; , \qquad x \in A \; ,$$

then f and A satisfy the inequality

$$\int_A f + \int_{\bar A} f^+ \geq 0 \; , \qquad \bar A = X - A \; .$$

PROOF. Suppose first that the sup in the hypothesis is > 0 in A. For any given $\epsilon > 0$, lemma 1 may then be applied. On using δ), α)

E.Hopf

and γ) we infer that

$$- \epsilon - \int_{\overline{A}} f^- < \int_A h^- - \int_{-A} h^- = -\int h^- \leq \int h \leq \int f = \int_A f + \int_{\overline{A}} f$$

and hence that (with $\epsilon > 0$ arbitrary)

$$\int_A f + \int_{\overline{A}} f^+ > - \epsilon .$$

The lemma is hereby proved under the stronger hypothesis, sup > 0 in A. Under the original hypothesis, sup ≥ 0 in A, we need merely use a function $p \in L$, $p > 0$ and apply the foregoing result to $f + \eta p$, $\eta > 0$, in place of f. The existence, by the way, of such a function p follows from the assumed σ-finiteness of the basic measure μ. The proof of lemma 2 becomes complete if we let $\eta \to 0$.

REMARK. It should be noted that lemma 2 is completely equivalent to the "maximal ergodic theorem",

$$(37) \qquad \int_B f \geq 0 , \qquad B = [x \mid \sup_{n>0} S_n f(x) \geq 0] ,$$

established by us previously and used in our proof of the ergodic theorem for a Markov operator with a finite invariant measure [14]. The equivalence becomes clear if three simple facts are noted. Firstly, the hypothesis of lemma 2 says that $A \subset B$. Secondly, the left hand side of the inequality of this lemma can be written (use that $B = B-A+A$, $\overline{A} = \overline{B} + B - A$)

$$\int_B f + \int_{B-A} f^- + \int_{\overline{B}} f^+ \geq \int_B f .$$

Thirdly, in $\overline{B}$ there holds sup $S_n f < 0$ and, in particular, $f = S_1 f < 0$ or, in other words, $f^+ = 0$. (37) obviously holds also if B is the set in which sup $S_n f > 0$. Lemma 1 is thus seen to furnish a simple proof of (37). However, we have mentioned lemma 2 because

E.Hopf

it is a handier form of (37).

The following lemma is a known consequence of (37) but it is more easily derived from lemma 2. It deals with the quotients

$$(38) \qquad q_n(x) = q_n(x;\ f,\ p) = \frac{S_n f(x)}{S_n p(x)}\ , \qquad n > 0\ .$$

Note that, for any constant ω,

$$(39) \qquad q_n(x;\ f - \omega p,\ p) = q_n(x;\ f,\ p) \doteq \omega\ .$$

LEMMA 3. If $f \in L$, $p \in L$, $p \geq 0$ then

$$\sup_{n>0} \left| q_n(x;\ f,\ p) \right| < \infty$$

holds almost everywhere in the set where $p > 0$.

PROOF. As $\left| q_n(x;\ f,\ p) \right| \leq q_n(x;\ |f|,\ p)$ holds it suffices to prove the lemma for the case that $f \geq 0$, and without the absolute value sign. Let

$$A = [x\,|\,p(x) > 0,\ \sup_{n>0} q_n(x;\ f,\ p) = \infty]\ .$$

By virtue of (39)

$$\sup_{n>0} q_n(x;\ g - \omega p,\ p) > 0$$

holds everywhere in A, for any constant $\omega > 0$. In other words

$$\sup_{n>0} S_n f'(x) > 0\ , \qquad f' = f - \omega p\ ,$$

must hold in A. Hence, by virtue of lemma 2,

$$\int_A (f - \omega p) + \int_{\bar{A}} (f - \omega p)^+ \geq 0\ .$$

From $f - \omega p < f$, $f \geq 0$ it follows that $(f - \omega p)^+ \leq f$. Therefore

$$\omega \int_A p \leq \int_A f + \int_{\bar{A}} f = \int f\ .$$

E.Hopf

As $\omega > 0$ is arbitrary and as $p > 0$ in A it follows that A has measure zero, and lemma 3 is proved.

For the following lemma of Chacon and Ornstein we give an abbreviated proof.

LEMMA 4. If $f \in L$, $p \in L$, $p \geq 0$, then

$$\lim_{n \to \infty} \frac{T^n f(x)}{S_n p(x)} = 0$$

holds almost everywhere in the set where $p > 0$.

PROOF. It sufficies to prove this for the case that $f \geq 0$. Pick a number $\epsilon > 0$ arbitrarily and consider the functions

$$g_n = T^n f - \epsilon S_n p , \qquad g_0 = f$$

which satisfy the relation

$$(40) \qquad\qquad g_{n+1} + \epsilon p = T g_n$$

since $T S_n = S_{n+1} - I$, $I =$ identity. We need merely show that, in almost every point of the set $[p > 0]$, $g_n < 0$ holds for all sufficiently large n or, in other words, that Σe_n converges where e_n is the characteristic function of the set $[g_n \geq 0]$. Using the relation $e_{n+1} g_{n+1} = g_{n+1}^+$ we infer from (40) and (33) that

$$\int g_{n+1}^+ + \epsilon \int e_{n+1} p = \int e_{n+1}(g_{n+1} + \epsilon p)$$

$$= \int e_{n+1} T g_n \leq \int e_{n+1} T g_n^+ \leq \int T g_n^+ \leq \int g_n^+ .$$

On adding up the resulting inequalities from $n = 0$ on ,

$$\int g_n^+ + \epsilon \int p \sum_1^n e_i \leq \int g_0^+$$

and hence

$$\int p \sum_1^\infty e_i \leq \epsilon^{-1} \int g_0^+ .$$

E. Hopf

$$\int p \sum_{1}^{\infty} e_i \le \epsilon^{-1} \int g_o^{+} \ .$$

Lemma 4 is herewith proved.

The next lemma contains the principal part of Chacon's and Ornstein's proof of the ergodic theorem.

LEMMA 5. Let $f \in L$; $p \in L$, $p \ge 0$. If in each point of a set A there holds

$$p > 0 \ , \quad \sum_{o}^{\infty} T^i p = \infty$$

and

$$\underline{\lim}_n q_n(x; f, p) < a < b < \overline{\lim} \ q_n \ (x; \ f, \ p)$$

where a, b are constants then A has measure zero.

PROOF. Write $q_n(f)$ for the quotients q_n and note that, in consequence of part of the hypothesis,

$$\sup_{n>0} q_n(f - bp) \ge \overline{\lim} \ q_n(f - bp) > 0$$

holds in the set A. Choose $\epsilon > 0$ arbitrarily and apply lemma 1 to f-bp in place of f. Write the h of this lemma in the form h-bp. We infer from a) that

$$(41) \qquad (h - b_p)^- \le (f - b_p)^-$$

and from δ) that

$$(42) \qquad \int_A (h - b_p)^- < \epsilon$$

holds. (41) and a < b imply that

$$(43) \qquad (h - ap)^- \le (f - ap)^-$$

holds a fortiori. In fact, (41) is equivalent to the statement that $h \ge f$ if $h < bp$. Therefore $h \ge f$ if $h < ap$, and this means that (43) holds. Now use β) of lemma 1, $h = f + Tg - g$,

E.Hopf

$$S_n h = S_n f + T^n g - g,$$

$$q_n(h) = q_n(f) + (T^n g - g)/S_n p \ .$$

On applying lemma 4 to g, p and on using the hypothesis that $S_\infty p = \infty$ on A we may infer that $q_n(h)$ and $q_n(f)$ have the same upper and the same lower limit (as $n \to \infty$) in each point of A except in a nulset which may safely be neglected in what follows. In particular, there holds

$$\underline{\lim} \ q_n(h) < a \quad \text{or} \quad \overline{\lim} \ q_n(ap-h) > 0$$

within A. Consequently, lemma 1 may be applied to ap-h in place of the f there. Write the new auxiliary function h of this lemma in the form ap-f'. Then it follows from γ) that

$$(44) \qquad \int(ap-f') \leq \int(ap-h) \quad \text{or} \quad \int(f'-ap) \geq \int(h-ap)$$

and from α) that

$$(45) \qquad (ap-f')^- \leq (ap-h)^- \quad \text{or} \quad (f'-ap)^+ \leq (h-ap)^+$$

and, finally, from δ) that

$$(46) \qquad \int_A (ap-f')^- = \int_A (f'-ap)^+ < \epsilon \ .$$

We now have a set of inequalities involving beside the given f two auxiliary h, f'. h can be eliminated from them in the following way. Split the integrands in the second inequality (44) into positive and negative parts,

$$(47) \qquad \int(f'-ap)^- \leq \int(h-ap)^- + \int\{(f'-ap)^+ - (h-ap)^+\} \ .$$

By virtue of (45) the second integrand on the right is ≤ 0. (47), therefore, stays valid if the integral is taken over A.

E.Hopf

Now,

$$(h-ap)^+ \geq h-ap = h-bp + (b\ a)p \geq - (h-bp)^- + (b-a)p \ .$$

Hence the right hand side in (47) is $\leq$

$$\int (h-ap)^- + \int_A (f'-ap)^+ + \int_A (h-bp)^- - (b-a) \int_A p \ .$$

On applying (43) to the first term, (46) to the second and (42) to the third term we obtain that

$$(48) \qquad \int (f'-ap)^- \leq \int (f-ap) + 2\epsilon -(b-a) \int_A p \ .$$

By the same reasoning as above (the use of β) of lemma 1) it follows that $q_n(f')$ and $q_n(h)$ have the same upper and the same lower limit in almost each point of A. Thereby the following result is obtained. If f satisfies the hypothesis of lemma 5 then, to any given $\epsilon > 0$, there exists a function f' which satisfies the same hypothesis and, in addition, the inequality (48). This result may in turn be applied to f' and we get a function f" which satisfies the hypothesis of the lemma and (48), with f, f' replaced by f', f", and so forth. On writing these inequalities underneath each other and on adding up the first n of them we find, on neglecting the non-negative integral remaining on the left, that

$$n \ \{(b-a) \int_A p - 2\epsilon\} \leq \int (f-ap)$$

must hold for any integer n. Hence

$$(b-a) \int_A p \leq 2\epsilon \ .$$

Lemma 5 follows from this since ϵ was arbitrary and since $a < b$, and $p > 0$ in A.

E.Hopf

COMPLETION OF THE PROOF OF THE SECOND ERGODIC THEOREM. We prove first that the q_n have a finite limit almost everywhere in the set $[p > 0]$. We may suppose that $f \geq 0$. Lemma 3 implies that the finite or infinite limit if it exists must be finite almost everywhere. So we need only prove existence. Within the part of the set $[p > 0]$ in which $\sum_o^\infty T^i p < \infty$ this is obvious because lemma 3 then implies that the series $\Sigma T^i p$ with non-negative terms has a finite sum almost everywhere in this part. In the remaining part $[\sum_o^\infty T^i p = \infty]$ of $[p > 0]$, however, the positivity of the measure of the set in which $\underline{\lim} < \overline{\lim}$ holds would imply the existence of two rational numbers $a < b$ such that the set $[\underline{\lim} < a < b < \overline{\lim}]$ would have positive measure. This would contradict lemma 5. Consequently, $\underline{\lim} = \overline{\lim}$ must hold almost everywhere in that remaining part.

Now we can easily prove the full theorem: The q_n have a finite limit almost everywhere in the set $[\Sigma T^i p > 0]$. Observe that this set is the union of the sets $[p > 0]$ and

$$(49) \qquad [S_{k+1} p > 0, \; S_k p = 0] \; , \; k > 0 \; .$$

In the latter set, $T^k p = S_{k+1} p - S_k p > 0$. On using two new functions

$$f' = T^k f \; , \; p' = T^k p$$

we find that in that set (49) and for $n > k$

$$\frac{S_{n-k} f'}{S_{n-k} p'} = \frac{S_n f - S_k f}{S_n p} = \frac{S_n f}{S_n p} - \frac{S_k f}{S_n p} \; .$$

From this formula and from the previous result as applied to f', p' in (49) it follows now that the q_n have a finite limit almost everywhere in the set (49). The second ergodic theorem is hereby completely proved.

E. Hopf

1d. CONSERVATIVE AND DISSIPATIVE PART OF A MARKOV PROCESS.

Consider again a positive linear operator T from L_μ to L_μ with a norm $\|T\| = 1$. We now turn to the simpler question of convergence or divergence of the infinite series

$$S_\infty f = \sum_0^\infty T^i f \, , \qquad f \in L_\mu \, .$$

DECOMPOSITION. X is the union of two disjoint sets, the convergence set X_c and the divergence set X_d, with the following properties. If $p \in L_\mu$, $p \geq 0$, $p > 0$ in X_d, then $S_\infty p = \infty$ holds almost everywhere in X_d in the sense of the measure μ . For any $f \in L_\mu$ the series $S_\infty f$ converges absolutely almost everywhere in X_c.

PROOF. Let $q > 0$, $q \in L$ and denote by X_d and X_c, respectively, the sets where $S_\infty q = \infty$, $< \infty$. Consider a function p with the properties stated above and apply lemma 3 to the quotients $S_n q / S_n p$. As $p > 0$ and $S_\infty q = \infty$ holds in X_d it follows from this lemma that $S_\infty p = \infty$ must hold a.e. in X_d. Now let $f \in L_\mu$ and consider the function $p' = q + |f| \in L_\mu$, $p' > 0$. Apply the same lemma to the quotients $S_n p' / S_n q$. As $S_\infty q < \infty$ holds in X_c it follows that also $S_\infty p' < \infty$ holds a.e. in X_c. The same holds therefore for $S_\infty |f|$. The rest of the lemma follows from $|T^n f| \leq T^n |f|$.

This result shows that within X_c the ergodic theorem is trivial and uninteresting. Both denominator and numerator converge separately a.e. in X_c. It is possible to separate X_d from X_c in such a way that the operator T somehow splits into two operators one of which acts on and produces functions that vanish in X_c whereas the other does the same relative to X_d? In other words, is the implication

$$(50) \qquad f = 0 \ \text{a.e. in} \ X_c \implies Tf = 0 \ \text{a.e. in} \ X_c$$

E.Hopf

valid? And is the analogous implication valid relative to X_d?
(50) means that the values of Tg, for general g, within X_c do not
depend on the values of g within X_d. From the representation formula

$$\int_A Tf \; d\mu = \int_X P(x,A)fd\mu$$

we easily infer another equivalent of (50): $P(x, X_c) = 0$ holds a.e.
in X_d.

(50) is valid [14] and its proof is given below. However, the
other implication regarding X_d does not generally hold good. It is
certainly valid if T is generated by a one-to-one point transformation but in the other extreme case where T is a completely continuous
operator the results of Yoshida and Kakutani [20] show that it is
invalid or, in other words, that X_c has an influence on X_d.

PROOF OF (50). We may assume that

$$(51) \qquad\qquad f \geq 0 \; .$$

Consider a fixed $q > 0$, $q \in L_\mu$. By the decomposition theorem,

$$(52) \qquad\qquad S_\infty q = \infty \quad \text{a.e. in } X_d$$

Note that $S_n q = S_{n+1} q$ and that

$$(53) \qquad\qquad TS_n q = S_{n+1} q - q \leq S_\infty q \; .$$

On letting

$$(54) \qquad\qquad h_n(x) = \inf (f(x), S_n q(x))$$

we can infer from (51), (52) and the premise of (50) that

$$(55) \qquad\qquad 0 \leq h_n \uparrow f$$

E.Hopf

holds in X_c as well as in X_d. We show that a.e.

$$(56) \qquad\qquad Th_n \uparrow Tf .$$

In fact $h_n \leq h_{n+1} \leq f \Rightarrow Th_n \leq Th_{n+1} \leq Tf$ and, by virtue of (33),

$$\int_X (Tf - Th_n)d\mu = \int_X T(f - hn)d\mu \leq \int_X (f - hn)d\mu \downarrow 0 .$$

From (54) and (53), $Th_n \leq TS_n q \leq S_\infty q$ and from (56),

$$Tf \leq S_\infty q$$

almost everywhere. This holds for any f which satisfies the hypothesis and, therefore, also for $\epsilon^{-1}f$, $\epsilon > 0$,

$$Tf \leq \epsilon S_\infty q$$

almost everywhere. This holds for any f which satisfies the hypothesis and, therefore, also for $\epsilon^{-1}f$, $\epsilon > 0$,

$$Tf \leq \epsilon S_\infty q .$$

As $S_\infty q < \infty$ holds a.e. in X_c it now follows that Tf which is ≥ 0 must be $= 0$ a.e. in X_c. (50) is herewith proved.

In one special case the sets X_d and X_c can be characterized in another way. Suppose that T is induced by a one-to-one mapping $x' = Tx$ of X onto itself such that T as well as T^{-1} are measurable and map sets with $\mu = 0$ (μ given and σ-finite on $\mathfrak{F}$) again onto such sets. It was above that such a mapping may be regarded as a Markov operator T from L_μ to L_μ,

$$(57) \qquad Tf(x) = r(x).f(Tx)$$

where the $\mathfrak{F}$-measurable $r(x) > 0$ is essentially uniquely determined. The sets X_d, X_c are thereby well defined. We recall the familiar

E.Hopf

concept of a "wandering set" A with respect to the mapping: $A \in \mathfrak{F}$, the images $T^n A$, $n \gtreqless 0$, are all disjoint. The following known theorem is worth mentioning.

$$X_c = \bigcup_{-\infty}^{\infty} T^i A$$ where A is a wandering set. X_d contains no wandering set of positive measure μ.

The equality is, of course, understood to hold up to a set with $\mu = 0$. It follows, by the way, that X_c and $X_d = X - X_c$ are both invariant under the mapping. The theorem characterizes X_c as the essentially largest among all sets of the form $\bigcup_{-\infty}^{\infty} T^i B$, B wandering. A set of this form is usually called *"dissipative"* with respect to the mapping, or the mapping is called dissipative within this set. A set, however, which is invariant under the mapping and which contains no wandering set of positive measure is sometimes called *"conservative"* with respect to the mapping. Or the mapping is said to be conservative within this set. Adopting this nomenclature wa can say : *The mapping is dissipative within X_c and conservative within X_d.*

A classical fact : If μ is finite and invariant under the point transformation then $X = X_d$. If, however, μ is infinite and invariant $X = X_c$ is possible. Example : X = infinite line, T = translation, μ = Borel measure. If a continuous group T^t of mappings is considered instead of the discrete group T^n then there is the example of the geodetic flow on a complete surface of constant negative curvature of the second class which is a dissipative flow. An open problem : Is the phase motion in the classical problem of three bodies dissipative? This would imply that the maximal distance between the three bodies tends, in general, to infinity as $t \to \infty$.

Let us return, for a few moments, to the theorem stated above and to its proof. Only the barest outline of this proof is given here. We start with dissipative sets in X and bring in the sets X_c, X_d, later. The proof splits into the following steps. a) The union of two dissipative sets is again a dissipative set. b) If μ is not finite use an equivalent (same nulsets) finite measure μ', consider the supremum M of the measures $\mu'(D)$ of all dissipative sets $D \subset X$ and show that there exists a dissipative set D^* with $\mu'(D^*) = M$. c) $X-D^*$ is conservative. d) Show that, for any $p \in L_\mu$, $p > 0$, $S_\infty p < \infty$ holds a.e. in D^* . e) Show conversely that, if $p \in L_\mu$, $p > 0$ and if $S_\infty p < \infty$ holds in a set A, $\mu(A) > 0$, A must contain a wandering set of positive measure. It is clear that these five steps make up the complete proof of the theorem.

We merely prove d) and e). Recall that the associated operator on set functions is $T\phi(A) = \phi(TA)$, $A \in \mathfrak{F}$ and that

$$\pi(TA) \equiv \int_A Tp \, d\mu \quad if \quad \pi(B) \equiv \int_B pd\mu$$

where T is given by (57). The left hand identity holds for all powers T^n, $n \geq 0$, too. Hence

$$(58) \qquad \sum_0^\infty \pi(T^i A) = \int_A S_\infty p \, d\mu, \quad A \in \mathfrak{F} \quad .$$

To prove d) take for A a wandering set such that D^* is the union of all its images $T^i A$, $i \gtrless 0$. As these images are pairwise disjoint it follows that the left hand side of (58) is $\leq \pi(D^*) \leq \pi(X) < \infty$. Consequently, $S_\infty p$ is a.e. finite in A. Since, however, every set $T^n A$, $n \gtrless 0$, is wandering too the same must hold within this set and, therefore, in their union D^*. To prove e) observe that the set A mentioned there necessarily contains a set B, $0 < \mu(B) < \infty$, $\pi(B) > 0$, such that $S_\infty p$ is bounded in B. On applying (58) to B we infer that

$$(59) \qquad \sum_0^\infty \pi(T^i B) < \infty .$$

E.Hopf

Now let $C_n = \bigcup\limits_{n}^{\infty} T^i B$ and note that $B = C_o \supset C_1 \supset \ldots$, $\pi(C_o) > 0$. By virtue of (58),

$$\pi(C_n) \leq \sum\limits_{n}^{\infty} \pi(T^i B)$$

can be made arbitrarily small as n gets large, certainly $< \pi(C_o)$ for some $n = k$. Consequently, $C_o - C_k$ has positive measure μ . The same must be true for at least one set $A' = C_i - C_{i+1}$, $i = 0,\ldots,k-1$. Now observe that

$$T^m A' = C_{i+m} - C_{i+m+1} , \quad m \geq 0 ,$$

and that these sets, $m > 0$, are disjoint with A'. As is wellknown this (and the assumption that T is one-to-one) implies that A' is a wandering set. The proof of the theorem is thereby finished.

For the validity of the theorem the hypothesis that the mapping is one-to-one is absolutely essential. Recall again that the sets X_c, X_d are determined by a Markov operator in the first place. Now, if the mapping is one-to-one there is only one Markov operator which can be reasonably connected with the mapping, and so it is not surprising that X_c, X_d can be characterized by the mapping alone. The situation is entirely different in the case of a many-to-one mapping. We have seen above that, in general, the representation of such a mapping by a Markov operator is no longer unique. That this ambiguity has an influence on the sets X_c, X_d is clearly illustrated by the following example.

Let X be the set of all natural numbers, $x = 1, 2,\ldots$. Let μ be the σ-finite measure which is one for each such x. Consider the point mapping

$$Tx = \begin{cases} x-1, & x > 1 , \\ 1 & x = 1 , \end{cases}$$

E. Hopf

of X onto X. $\mathcal{F}$ consists of all subsets of X, $T^{-1}\mathcal{F}$ only of those subsets which contain either none of the two numbers 1, 2 or both. A Markov operator associated with T is of the form (57) where $r \geq 0$ is such that (26), the conservation property of Markov operators, is satisfied for each $f \in L_{\mu}$,

$$\sum_{-\infty}^{\infty} f(x) = \sum_{-\infty}^{\infty} r(x) \, f(Tx) \, .$$

This requirement is readily seen to be equivalent to the relations

$$r(1) + r(2) = 1 \, , \qquad r(3) = r(4) = \ldots = 1 \, .$$

So we have infinitely many Markov operators (57). There holds

$$(60) \qquad T^n f(x) = r_n(x) \, f(T^n x) \, , \qquad r_n = r(x) \, r(Tx) \ldots r(T^{n-1} x).$$

First case. $r(1) = 1$, $r(2) = 0$. We distinguish between the two parts

$$(61) \qquad\qquad [x \,|\, x = 1], \quad [x \,|\, x > 1]$$

of X. In the first set, $r_n = 1$, $S_{\infty} f = \infty \cdot f(1)$. However, in the second set, $r_n(x)$ contains, for all large n, the factor $r(2) = 0$. Therefore $S_{\infty} f$ is finite. It follows that the first set is X_d and the second X_c.

Second case. $r(1) < 1$. No matter what x we start from the factors of r_n are $= r(1)$ from a certain place on. Hence, $r_n(x) < C(x) r(1)^n$ and, consequently, $X_c = X$, $X_d = 0$. Note, by the way, that the unique r(x) which is $T^{-1}\mathcal{F}$ -measurable, $r(1) = r(2)$, comes under this second case.

It is instructive in this connection to consider another example in which the space X is not countable. This example was briefly mentioned already : x is an angle variable mod. 2π and the

E.Hopf

mapping is

$$x' = Tx = 2x \bmod 2\pi .$$

We found that each measurable function $r(x) \geq 0$ with satisfies the relation

$$r(x) + r(x + \pi) \equiv 2$$

furnishes a Markov operator T ,

$$Tf(x) = r(x).f(Tx) ,$$

for the mapping T. In order to study the behaviour of the iterates T^n of the operator we begin with the simplest case $r = 1$. It was already mentioned that, for $g \in L_\mu = L^1$, the limit

$$\lim_{n \to \infty} \frac{1}{n} \sum_{o}^{n-1} g(2^\nu x)$$

exists almost everywhere. We need the fact that the limit function $\tilde{g}(x)$ is constant a.e., and that the constant equals

$$\frac{1}{2\pi} \int_{o}^{2\pi} g(y)\, dy .$$

The simple reason for this is : For almost all x there holds $g(2x) = g(x)$, $g(2^k x) = g(x)$ where k is an arbitrary positive integer. The Fourier expansion of $\tilde{g}$ must therefore reduce to the constant term. Now we return to our question regarding the associated operator

$$Tf(x) = r(x).f(2x)$$

where $r(x)$ is any admissible function > 0. X is the set on the circle in which every series

$$S_\infty p(x) = \sum_{o}^{\infty} T^i p(x)$$

E.Hopf

$p > 0$, $p \in L^1$, converges a.e.. Now,

$$T^n p(x) = r_n(x) p(2^n x) \, , \quad r_n(x) = r(x) \ldots r(2^{n-1} x)$$

and $r_0(x) = 1$. From the immediately preceding result, $g = \log r$, it follows that a.e.

$$\log \sqrt[n]{r_n(x)} \ \rightarrow \ \frac{1}{2\pi} \int_0^{2\pi} \log r(y) dy \, .$$

As the log is convex the right hand constant is

$$\leq \log \left\{ \frac{1}{2\pi} \int_0^{2\pi} r(y) \, dy \right\}$$

and equal to it if and only if r is constant a.e.. The last named expression has the value zero as follows readily from the admissi- bility relation for $r(x)$. Hence we infer that, for almost every x,

$$\sqrt[n]{r_n(x)}$$

has a limit < 1 unless $r(x)$ is constant a.e. (which can happen only if $r = 1$ a.e.). If we combine this with the fact that the set X_c is essentially independent of the choice of the function p, $p > 0$, $p \in L^1$, (we may take $p = 1$, $T^n p = r_n$) we obtain the following cu- rious result.

Any admissible function r, $r > 0$, $r(x) + r(x + \pi) = 2$, furni- shes a Markov operator "associated" with the mapping. In the case where $r = 1$ a.e. we have $X_c = 0$ but if r is not equivalent to the function "one" then we have $X_c = X$!

We had noticed this fact already in 1946. There is, of course, the question wether there is an "optimal" Markov operator which de- scribes the future behaviour of the iterates of a many-to-one map- ping better than the other associate operators. On comparing the

E. Hopf

results of the two examples mentioned here we can see that a generally valid answer cannot be trivial. However, the beautiful applications of information theory to ergodic theory which Kolmogorov has initiated and to which we were introduced in this Seminar by Paul Halmos may throw light on this question.

Let us finally return to the general second ergodic theorem itself. We wish to conclude this section with the question (not answered as yet) about the nature of the limit function $\lim q_n(x)$. This question is, of course, interesting only in X_d.

If we operate in X_d only, more precisely, if we confine ourselves to functions f vanishing in X_c then (50) has the following implication. Within X_d, T represents again an operator of the same kind, positive, linear and of L_μ - norm ≤ 1 within X_d. Furthermore, if T is a Markov operator in X so is T in X_d. We may therefore assume, without loss of generality, that

$$X_d = X \iff S_\infty p = \infty \quad \text{in } X \text{ if } p > 0 .$$

We need the notion of the dual operator $\tilde{T}$ associated with the given operator T,

$$(D) \qquad \int g T f \, d\mu \equiv \int f \tilde{T} g \, d\mu ,$$

explicitely,

$$\tilde{T} g(x) = \int P(x, \, dy) g(y) .$$

$\tilde{T}$ is actually an operator from μ-equivalence class of bounded functions to again such a class [14]. $\tilde{T}$ is positive, and $\tilde{T} 1 = 1$ holds if T is a Markoff operator. We assume here that T is a Markoff operator (this property is used in the proof of one of the subsequently mentioned facts).

E.Hopf

A bounded and μ-measurable function h(x) is said to be invariant under the Markoff process if it is invariant under $\tilde{T}$,

(62) $$\tilde{T}h = h$$

(up to a set with $\mu = 0$). The theory of these invariant functions
was developed in [14], however, under restrction to bounded functions. The main results are the following (see [14], §9).

The invariant functions form an algebraic field. Also, $|h|$
is invariant if h is.

If h is invariant then there holds (almost everywhere)

(63) $$\tilde{T}(h\ w) = h\ \tilde{T}(w)$$

for every bounded and measurable function w(x). Conversely, of course, the latter property of h implies invariance of h since $\tilde{T}1 = 1$.
The proof of (63) is given in [14], §9 . The property (63) is completely equivalent to the following property of the original operator T :

(64) $$T(h.w') = h\ T(w')$$

holds for every μ-integrable function w'(x). The equivalence of
the two properties follows easily by means of the duality relation (D).

Let us now return to the question of the limit function

$$f^{*} = \lim_{n\to\infty} \frac{S_n f}{S_n p}$$

of the general ergodic theorem. It has not yet been proved that
this limit function is invariant and that it is independently characterized by the condition that pf^{*} is μ-integrable and that

$$\int hf^{*}pd\mu = \int hfd\mu$$

E.Hopf

holds for every bounded and invariant function h. We repeat that this conjecture is stated under the hypothesis that T is a Markov operator. In order to prove it is first of all necessary to extend the notion of invariant function from bounded functions to functions h' for which

$$\int |h'| \, pd\mu$$

is finite. This should not be difficult. We also would like to call attention to another way of proving the general ergodic theorem, namely, by generalizing a simple trick which F. Riesz once introduced to prove the classical mean ergodic theorem. Consider first functions of the form

$$Tg - g.$$

The functions h which are orthogonal to all of them must satisfy the relation

$$0 = \int (Tg - g) \, hd\mu \equiv \int g(\tilde{T}h - h)d\mu$$

for all g or, in other words, h must be invariant, $\tilde{T}h = h$. Show that every f can be approximated by a sum of a function $Tg - g$ and an invariant function h. Then prove the ergodic theorem in the two simple cases 1) $f = Tg - g$, 2) $f = h$. In the first case use lemma 4 (with g written in place of f) to show that $S_n(Tg - g)/S_n p \to 0$. In the second case use the relations

$$T^i(h \cdot p) = h \, T^i p$$

which follow by repeated application of (64). In this case there would hold

$$S_n h \, / \, S_n p = h$$

for all n. Finally, it must be shown that, for a function f' of

E.Hopf

small norm, the function

$$\limsup_{n\to\infty} \left| \frac{S_n f'}{S_n p} \right|$$

is also small except in a set of small measure. This can be inferred from our maximal ergodic lemma (lemma 2). This proof of the ergodic theorem, if carried through completely, would not only prove
the existence of the limit function but also its invariance.

E.Hopf

2a. TWODIMENSIONAL HYPERBOLIC GEOMETRY.

The well-known Poincaré model of the hyperbolic plane is the interior of the unit circle endowed with the metric

$$(1) \qquad ds^2 = 4 \frac{dx_1^2 + dx_2^2}{(1 - x_1^2 - x_2^2)^2}$$

which has curvature minus one. The isometries in this geometry are those Moebius transformations that map $x_1^2 + x_2^2 < 1$ onto itself. An isometry is completely determined by the requirement that it carry a given line-element (such elements are understood to be directed) into another such element. Of course, isometries leave all quantities unchanged which are determined by ds only, angles (= euclidean angles), element of area

$$(2) \qquad dA = \frac{4dx_1dx_2}{(1 - x_1^2 - x_2^2)^2} .$$

Geodesics are carried into geodesics. They are the arcs in $x_1^2 + x_2^2 < 1$ of the circles orthogonal to the unit circle. Hyperbolic distance between two points $x = (x_1, x_2)$, $x' = (x_1', x_2')$ is denoted by $s(x,x')$.

Consider now the threedimensional space of line-elements e in the hyperbolic plane. Each isometry of this plane induces, of course, a mapping of that space onto itself. We introduce as metric in the line-element space the expression

$$(3) \qquad d\sigma^2 = ds^2 + d\chi^2$$

where $d\chi$ is determined in the following way. Consider two nearby elements e, e' with bearer points x, x'. Move e from x to x' by parallel displacement, in other words, move the element to x' along the geodesic from x to x' in such a way that its direction makes always the same angle with the geodesic. $d\chi$ is then the angle between the element e in its final position and e'. Obviously,

E. Hopf

$|d\chi|$ is independent of the order of the two elements. $d\sigma$ is therefore a Riemannian metric in e-space. $d\sigma$ is even invariant under the
mappings induced by the isometries of the plane since the operation
of parallel displacement is invariant under isometries. In other
words, those mappings are themselves isometries in e-space relative to $d\sigma$. We denote by $\sigma(e, e')$ the invariant distance defined by
$d\sigma$ in e-space. The invariant element of volume-measure induced by
$d\sigma$ in e-space is found to be

$$(4) \qquad dm = d\phi dA$$

where $d\phi$ is the angle-differential in a point of the plane.

If we consider in e-space the motion along the geodesics with
speed $ds/dt = 1$ we obtain a one-parameter group of mappings $T^t e$,
$T^{t+s} = T^t T^s$, of e-space onto itself: $T^t e$ is the position attained
by e after it has moved t units along the geodesic determined by
it. This is the geodetic flow in the e-space of the hyperbolic plane. It is a well-known fact of differential geometry that the geodetic flow on a surface leaves the element of measure (4) in e-space invariant. The same is, of course, true about the Lebesgue measure m determined by it in e-space. m is invariant not only under
isometries but also under the geodetic flow.

We need the following simple coordinate-representation of the
geodesics and the flow along them. We assume throughout that a "geodesic" is a directed geodesic. A geodesic is characterized by its
points of infinity θ^-, θ^+ on the unit circle, $\theta^- \neq \theta^+$, where θ is
an angle (mod 2π). And a line-element e is characterized by the geo-
desic determined by it and by the position of its bearer point on
this geodesic (orthogonal circular arc). This position is, in turn,
characterized by the distance $s \gtreqless 0$ of the point from the euclidean

E.Hopf

midpoint of this directed arc,

$$(5) \qquad\qquad e = (\theta^-,\ \theta^+,\ s).$$

The geodetic flow is then simply

$$(6) \qquad\qquad T^t e = (\theta^-,\ \theta^+,\ s+t).$$

The invariant measure-differential dm in e-space has the simple form

$$(7) \qquad\qquad dm = \rho(\theta^-,\ \theta^+)\ d\theta^- d\theta^+ ds\ .$$

That ρ is independent of s follows from the invariance of dm under T^t. Positivity and continuity is the only property of this function ρ which is needed in what follows.

We also need the following fundamental fact about hyperbolic geometry: To every geodesic and to every point x there exists precisely one geodesic through x that has the same point plus infinity on the unit circle. Two geodesics with the same plus-infinite point we call *positively asymptotic* (to each other). The asymptotic character is expressed by the following.

LEMMA 1. If two line-elements e, e' determine positively asymptotic geodesics then there exists a number a (depending on e, e') such that

$$\sigma(T^{t+a} e,\ T^t e') \to 0 \qquad as \qquad t \to \infty\ .$$

This is most easily proved if we map the interior of $x_1^2 + x_2^2 = 1$ by a Moebius-transformation onto the upper half plane of a y-plane, $y = (y_1,\ y_2)$. We do this in such a way that the common point $+\infty$ of the two geodesics goes into $y = \infty$. The two geodesics become two straight lines l, l' orthogonal to the line $y_2 = 0$. Consider the

two tangent elements e on l, e' on l'. Evidently the desired value
of a is the one for which T^ae and e' have the same coordinate y_2.
It is geometrically obvious that the non-euclidean distance $\int ds$
between the bearer points of the elements T^{t+a}e, T^te' tends to zero
as $t \to \infty$. At the same time the geodesic arc joining them becomes
straighter and straighter in the euclidean sense. Now, the distance

$$\sigma(T^{t+a}e,\ T^t e')$$

is not greater than the length of any path in e-space joining the
two elements. Choose for this path the geodesic arc between the bea
rer points plus parallel displacement of the first element along
it $(d\sigma = ds)$, and then the remaining rotation of the element into
the second element $(d\sigma = |d\chi|)$. This makes the lemma obvious.

2b. COMPLETE SURFACES γ OF CONSTANT NEGATIVE CURVATURE.

It is a classical fact that every abstract twodimensional Riemannian manifold of curvature minus one which is complete (P.Koebe,
H.Hopf and W.Rhinow, every geodetic arc can be continued on the manifold to all values of the length-parameter s) can be realized as
follows. Let G denote a group of isometries S . We suppose that it
is discrete or, in other words, that it does not contain isometries
arbitrarily close to the identity. Well-known consequences of the
discreteness are : 1) G is countable, 2) the set of distinct points
Sx, S $\in$ G congruent to a point x within the unit circle does not
have a cluster point within this circle, 3) the fixed points of the
isometries S $\in$ G different from the identity form a countable set
of single points and orthogonal arcs. G possesses a fundamental domain D within the unit circle. The classical construction of a simple fundamental domain is this. Choose a point x^o within the unit

E.Hopf

circle which is no fixed point of any S $\in$ G different from the i-
dentity. Then the set of all points x which satisfy the inequalities

$$s(x, \; x^o) < s(x, \; Sx^o)$$

for all S $\in$ G different from the identity is a fundamental domain
D = D(x^o) for G . D is geodetically convex. Its boundary is formed
by countably many geodetic arcs and, perhaps, parts of the unit circle itself. This is so because the equation $s(x,x^o) = s(x,x^1)$ defines a geodesic in hyperbolic geometry. Fixed points can obviously
occur only on the boundary of such a D .

   We obtain a Riemannian surface γ of curvature minus one if we
identify all points congruent to a given point x and if we regard
the set

$$p = [Sx \mid S \in G]$$

as a single point p . Distance between such a point and another
such point

$$p' = [Sx' \mid S \in G]$$

is defined by

$$s(p,p') = \inf_{S \in G, S' \in G} s(Sx, \; S'x') = \inf_{S \in G} s(Sx, x') = \inf_{S' \in G} s(x, S'x').$$

The last two equations hold by virtue of the invariance of s . Si-
milarly, the directed line-elements P on γ are defined by identifi-
cation of congruent line-elements on γ , and distance of two such
elements P is defined by

$$\sigma(P, \; P') = \inf_{S \in G, S' \in G} \sigma(Se, \; S'e')$$

where S, S' mean the isometries induced in e-space and G means
the corresponding group. The threedimensional space of line-ele-

E.Hopf

ments P on γ is denoted by Ω . All the other quantities which are defined within the unit circle and invariant under isometries define similar quantities on γ . Plane measurability of a set of points p on γ means measurability of the set of *all* representants x in $x_1^2 + x_2^2 < 1$ of all those p . A set in γ is said to have measure zero if that set of all representant points has this property. Of course, measure of a general measurable set on γ has to be defined somewhat more carefully. It is defined as the measure $\int dA$ of the intersection of that set of all representants x with a fundamental domain D of G . It is an easy consequence of the invariance of $\int dA$ under isometries that the measure defined in this way is independent of the particular D . Measure zero, according to this definition, agrees with measure zero as defined a moment ago because the set of all representants x is the union of all copies under G of those intersections and because these copies are countable in number. In a perfectly analogous way m-measurability and measure m are defined in the space Ω of line-elements P on γ .

The geodetic flow $T^t P$ is unambiguously defined on γ or, rather, in Ω because

$$ST^t e = T^t Se$$

holds for all induced isometries S . The measure m in Ω is invariant under T^t . A geodesic on γ is said to be positively asymptotic to another geodesic on γ if some representant of the first is pos. asymptotic to some representant of the second in the unit circle. Two streamlines $T^t P$, $T^t P'$ are said to be pos. asymptotic if the geodesics represented by them on γ have this property.

LEMMA 2. If the streamlines $T^t P$, $T^t P'$ through two points P, P' of Ω are pos. asymptotic then there exists a number a such that

E. Hopf

$$\sigma(T^{t+a} P, \, T^{t} P') \to 0 \quad \text{as} \quad t \to \infty .$$

This is an immediate consequence of lemma 1 and of the fact that $\sigma(Q, \, Q') \leq \sigma(e, \, e')$ holds for any representative $e, \, e'$ of two points $Q, \, Q'$ of Ω .

We add here a little remark which has to be used later.

LEMMA 3. Let Δ be a set of directions in a point p of a surface γ . Consider a second point p' on γ and a geodesic through p' which is pos. asymptotic to some geodesic passing through p in a direction of Δ . Denote by Δ' the set of directions in p' of all these geodesics through p'. Then, if Δ has angular measure zero so does Δ'.

To prove this we consider first two points $x, \, x'$ in the covering surface $x_1^2 + x_2^2 = 1$. Any geodesic through x determines a unique geodesic through x' which is pos. asymptotic to the first. In this way a one-to-one correspondence is established between the directions in x and those in x'. It is bi-analytic and, therefore, it carries a set of directions in x of angular measure zero into a set of the same sort in x'. Now consider two points $p, \, p'$ on the surface γ . There are, in general, many geodesics through p' pos. asymptotic on γ to a given geodesic through p. Pos. asymptoticity on γ means that, in the covering surface, some representant of the second geodesic is pos. asymptotic to some representant of the first. This implies that, on γ, the directions in p' corresponding to any given direction in p are furnished by countably many one-to-one correspondences of the kind mentioned above. The lemma is thereby proved.

2o. THE TWO CLASSES OF SURFACES γ.

Consider a geodesic g on a surface γ . We call it *positively divergent* on γ if, for $t \to \infty$, $s(p_t, p^o) \to \infty$ where p_t is the point

E.Hopf

running along g, more exactly, the p-coordinate of $T^t P$ and where p^o is an arbitrary fixed point on γ . Of course, if the statement holds true for one p^o then it does so for any other fixed p^o. Quite analogously, a streamline $T^t P$ in the P-space Ω over γ is called pos. divergent if, for $t \to \infty$, $\sigma(T^t P, P^o) \to \infty$ where P^o is an arbitrary fixed point in Ω . It is easily seen that both concepts mean the same thing for a geodesic in γ and the corresponding streamline in Ω . In fact, this follows from the general inequalities

$$(8) \qquad s(p,p') \le \sigma(P,P') \le s(p,p') + \pi$$

where p, p' are the bearer points of P, P', respectively. The second follows by parallel displacement of P along the geodesic arc from p to p' and by a rotation of the line-element in p'.

An important consequence of lemma 2 : If a geodesic on γ (streamline in Ω) is positively divergent in γ (Ω) so is every other geodesic (streamline) which is positively asymptotic to the first. This statement remains, of course, valid if the word "positively" is replaced by the word "negatively" throughout ($t \to -\infty$).

Lemma 3 in combination with this last statement permits the use of the abundance of divergent geodesics for a subdivision of the class of surfaces γ into two subclasses.

DEFINITION. A surface γ is of the first class if the divergent geodesics issuing from a point p of γ form a set of directions in p of angular measure zero. If this is true for one point of γ it is true for every other point of γ . γ is of the second class if it is not of first class.

The possible surfaces γ of first class represent, topologically, a vast variety of surfaces. There are many types of closed surfaces. For instance, every closed orientable surface of genus

98

F.Hopf

> 1 occurs among them. A closed surface γ is of first class because, obviously, no divergent geodesic exists on such a surface. More generally, the surface is of the first class if the group G of covering transformations possesses a fundamental domain D the closure of which is entirely in $x_1^2 + x_2^2 < 1$. Many surfaces γ with boundary fall into this category. In this case the problem of the geodesics is a non-euclidean billiard problem with reflection at the boundary.

Take, for example, an equilateral non-euclidean triangle with the three interior angles equal to $2\pi/2n = \pi/n$ where n is an integer > 3 (the sum of the three angles must be < π) and take the three hyperbolic reflections at the three sides as generators of a group G . The condition on the angle insures that the images under G of the triangular domain cover the hyperbolic plane simply or, in other words, that this triangle is actually a fundamental domain for this group G . The surface γ generated by G is then this triangular area, and the geodesics problem is the hyperbolic billiard problem.

Quite generally, we can say that a surface γ is of first class if its area A is finite. The reason is this. If γ has finite area its P-space Ω has finite volume $m(\Omega) = 2\pi A$. By virtue of Poincaré's recurrence theorem, almost all streamlines of the geodesic flow in Ω (with invariant measure m) are recurrent. However, if γ were of second class then, by virtue of the theorem of the next paragraph, almost every streamline would be divergent which is a contradiction. There are many types of surfaces γ of finite area but of infinite non-euclidean diameter. A well-known example is that of the modular group G which has a geodetic triangle for fundamental domain with one corner or, rather, cusp at infinity (on $x_1^2 + x_2^2 = 1$). The cusp

E.Hopf

does not prevent the domain from having a finite area.

A surface γ of second class is realized if the fundamental domain D of the group G has on its boundary an open arc of the unit circle. Any geodetic ray that ends up on this arc must be divergent on γ . In fact, such a ray stays ultimately in D and the point running along it gets farther and farther removed, in the sense of the distance s, from the finite part of the boundary of D. Consequently, its minimal distance from all the points Sx^o, $S \neq$ identity, congruent to a fixed point x^o inside D (they are outside of D) tends a fortiori to infinity. Obviously, the initial directions of the geodetic rays issuing from a fixed point and ending on that arc fill a whole angle. To the arc on the unit circle there corresponds an infinite funnel of the surface γ generated (at least this is true if γ has no finite boundary points). However, surfaces of second class can be much more complicated in that they can reach to infinity in much more intricate ways.

We mention that the classical two "kinds" of surfaces are not the same as our two "classes". "kind" is a purely topological and, as such, extremely natural motion. From the standpoint of ergodic theory, however, the natural division is the one into the two "classes". A more detailed discussion of the relation between "kinds" and "classes" can be found in our memoir [11].

2d. ERGODIC THEORY AND THE TWO CLASSES OF SURFACES γ .

We are now ready to begin with the proof of the two principal theorems on the geodetic flow on surfaces γ .

FIRST THEOREM. For a surface γ of first class the geodetic flow is ergodic. In other words, if f(P) and g(P) > 0 are m-integrable in Ω then

E.Hopf

$$\lim_{\tau \to \infty} \frac{\int_0^\tau f(T^t P)\,dt}{\int_0^\tau g(T^t P)\,dt} = \frac{\int_\Omega f\,dm}{\int_\Omega g\,dm}$$

holds for almost every $P \in \Omega$ in the sense of the measure m. The same holds for the limit as $\tau \to -\infty$.

SECOND THEOREM. For a surface γ of second class the geodetic flow is dissipative or, in other words, for almost every $P \in \Omega$ in the sense of the measure m the streamline $T^t P$ is divergent in Ω , positively as well as negatively.

Both these theorems have a common root, namely, the existence of asymptotic geodesics and the relations between them peculiar to hyperbolic geometry. Their effect is expressed in the following.

PRINCIPAL LEMMA. Let B_+ and B_- be two m-measurable sets in the line-element space Ω of a surface γ . Suppose that they satisfy the following conditions. a) Each set is invariant under the geodetic flow. b) With every streamline in B_+ every streamline pos. asymptotic to it is in B_+; the same holds for B_- with respect to neg. asymptoticity. c) The set of all points P in one set but not in the other has measure m = 0. Then, under these conditions, either B_+, B_- have both measure m = 0 or their complements in Ω have both measure m = 0.

First we prove this lemma and then the two main theorems. It is interesting to observe that the lemma is true if in the hypothesis c) and in the conclusion the statement "set of measure m = 0" is replaced by the statement "empty set". In this altered form the lemma is not only true but almost trivial. In fact, c) means then that $B_+ = B_-$, and the conclusion says that either $B_+ = B_- = 0$ or $B_+ = B_- = \Omega$ This modified lemma is true simply because to two arbitrary geodesics on γ there always exists at least one geodesic on γ which is neg. asymptotic to the first and pos. asymptotic to

the second (on the covering surface there is precisely one such
geodesic). If $B_+ = B_-$ is non-void then this argument can be applied
to an entirely arbitrary streamline and to a streamline in B^+ . However, we need the lemma in the form stated previously because, in
both our applications, the exceptional nulsets in the hypothesis c)
are, in general, non-void.

The proof of the principal lemma rests upon the same simple argument but then the exceptional sets of measure zero require care-
ful handling. A set of complete streamlines in the element-space
Ω of a surface γ can be represented in two ways, as a flow-invariant
point set in Ω and as set of pairs of points (θ^-, θ^+) on the unit
circle which is invariant under the simultaneous transformations
$S \in G$. Remember that m is the flow-invariant measure in Ω and that
$\iint d\theta^- d\theta^+$ is a measure in the space of those pairs (θ^-, θ^+) . We prove first: A flow-invariant point set in Ω has measure m = 0 if and
only if the corresponding set of pairs of points (θ^-, θ^+) has measure $\iint d\theta^- d\theta^+ = 0$. To see this denote by C the invariant set in
Ω . Let C' be the set of all elements e in the covering surface
$x_1^2 + x_2^2 < 1$ which represent the elements contained in C. We know
that m(C) = 0 implies m(C') = 0 and vice versa. C' is invariant under the geodetic flow in the covering surface. On using the coordi-
nates (θ^-, θ^+, s) in (6) for the elements of C' we see that C' is
a cylindrical set, with its base in (θ^-, θ^+)-space equal to that
set of pairs of points. By virtue of (7), m(C') = 0 means that that
set has measure $\iint \rho d\theta^- d\theta^+ = 0$. As ρ is > 0 and continuous this
means that that set has measure $\iint d\theta^- d\theta^+ = 0$. The statement is thereby proved.

The principal lemma is now proved as follows.
We assume that

E.Hopf

$$(9) \qquad m(B_-) > 0$$

and show that $\bar{B}_- = \Omega - B_-$ and $\bar{B}_+ = \Omega - B_+$ have measure $m = 0$. Retaining the letters B_-, B_+ for the sets corresponding to these sets in (θ^-, θ^+)-space we infer from the hypothesis b) of the lemma that both those sets are product sets in that space

$$(10) \qquad B_- = b_- \times \Theta , \quad B_+ = \Theta \times b_+$$

where Θ is the θ-line and where b_-, b_+ are certain measurable subsets of it. Strictly, we would have to exempt the diagonal-line $\theta^- = \theta^+$ from $\Theta \times \Theta$ since no geodesics correspond to the points on this line. Since, however, it is of measure $\iint d\theta^+ d\theta^+ = 0$ it may safely be disregarded in this proof. Hypothesis c) says that

$$(11) \qquad B_- \bullet \bar{B}_+ = b_- \times \bar{b}_+ , \quad \bar{B}_- \bullet B_+ = \bar{b}_- \times b_+$$

have measure zero. This may be understood in the sense of the product measure $\iint d\theta^- d\theta^+$. By virtue of (9), b_- has non-zero measure on the θ-line. Consequently, since the first set has product measure zero, $\bar{b}_+$ has θ-measure zero. Since the second set has product measure zero it now follows that $\bar{b}_-$ has θ-measure zero. In other words, the complements of both sets (10) have measure zero. These complements in product-space correspond exactly to the sets $\bar{B}_-$, $\bar{B}_+$ in the space Ω . Consequently, these latter sets have measure $m = 0$. The lemma is herewith completely proved.

We prove the second theorem first. Denote by $B_-(B_+)$ the set of all elements $P \in \Omega$ on neg. (pos.) divergent streamlines. As γ is of second class the angular measure $\int d\theta$ of the "pos. divergent" directions in a point $p \in \gamma$ is positive. Hence

$$m(B_+) = \iint_{B_+} d\phi dA = \int_{\gamma} \{\int d\phi\} dA > 0 .$$

E.Hopf

That the two sets B_- and B_+ satisfy hypothesis c) of the principal lemma follows from a general theorem ([9] or [10]) :

If $T^t P$, $T^o P = P$, $T^{t+s} = T^t T^s$, is a continuous flow in a complete metric space Ω with invariant measure m (m σ-finite) then the set of all streamlines that are divergent in one direction but not the other has measure m = 0.

That hypothesis b) is satisfied was remarked before. Therefore, the conclusion of the principal lemma holds, and the second theorem is thereby proved. By virtue of lemma 3 we have obtained a sharper result :

If γ is of second class then, in any point $p \in \gamma$, the geodesic rays issuing from p are divergent for almost all initial directions.

We now prove the first theorem. Just as in the beginning of the preceding proof we infer that, for a surface γ of first class, the set of all pos. or neg. divergent streamlines in Ω is a set of measure m = 0. We have to mention now that the general theorem referred to in the preceding proof is part of the following general theorem ([9] or [10]) :

Under the same hypothesis as in that theorem Ω splits into two invariant parts, the conservative and the dissipative part. The first contains almost no pos. or neg. divergent streamlines. The second consists almost exclusively of pos. as well as neg. divergent streamlines. In the conservative part

$$(12) \qquad \int_o^\infty g(T^t P)dt = \infty$$

holds for any g(P) > 0 almost everywhere.

In the first chapter of these lectures this decomposition was mentioned already for the case of a single mapping T (however, without the part about divergent streamlines).

E.Hopf

In our present case, geodetic flow on a surface γ of first class, the flow is purely conservative. For a conservative flow with invariant measure m the ergodic theorem states this [10] : If $f(P)$, $g(P) > 0$ belong to L_m (m-integrable) in Ω then

$$(13) \qquad \lim_{\tau \to \infty} q_\tau(P) = f^*(P) \ , \quad q_\tau(P) = \frac{\int_0^\tau f(T^t P)\,dt}{\int_0^\tau g(T^t P)\,dt}$$

exists almost everywhere (m) in Ω ; $gf^* \in L_m$, f^* is T^t-invariant and satisfies the relation

$$(14) \qquad \int_\Omega gf^* h\,dm = \int_\Omega fh\,dm$$

for every bounded and measurable $h(P)$ that is invariant under T^t. The average in the past,

$$(15) \qquad \lim_{\tau \to -\infty} q_\tau(P) = f^{**}(P)$$

exists almost everywhere too (apply theorem to the flow $\tilde{T}^t = T^{-t}$ which has the same invariant functions as T^t), and f^{**} satisfies the same relation (14) as f^*. Consequently,

$$\int_\Omega g(f^{**} - f^*)\,dm = 0$$

must hold for every bounded invariant h, so for instance, for $h = \mathrm{sign}\,(f^{**} - f^*)$. Hence,

$$(16) \qquad f^{**}(P) = f^*(P) \qquad a.e.$$

must hold.

Our aim is to show that $f^* = f^{**}$ is constant a.e.. From (14) $h = 1$, it would then follow that this constant has the value $\int f / \int g$. To prove constancy of f^* for any $f \in L_m$ it suffices to prove this for every f in a set of f's that is dense in L_m. Reason : The linear operator $f^* = T^* f$ (g is kept fixed) satisfies

E. Hopf

$$\int_{\Omega} g \, |f^{*}| \, dm \le \int_{\Omega} |f| \, dm$$

(apply (14) with h = sign (f^{*})). We use this fact as follows in
our present case (γ, Ω). We choose the fixed function g > 0 such
that

$$(17) \qquad \frac{g(P') - g(P)}{g(P')} \; \rightrightarrows \; 0 \qquad \text{as} \quad \sigma(P,P') \to 0$$

holds *uniformly* with respect to P, P'. In the important special case where γ has finite area we may simply choose g = 1. For functions
f we take only those f's which have the similar property that

$$(18) \qquad \frac{f(P') - f(P)}{g(P')} \; \to \; 0 \qquad \text{as} \qquad \sigma(P,P') \to 0$$

holds *uniformly* in Ω . In the case where γ has finite area and where g = 1 this simply means that f is uniformly continuous in Ω .
It can be shown by means of classical arguments that the set of those f's is dense in L_m.

After these preparations we now turn to be main point of our
proof. Consider g, f, as indicated above. We show first that

$$(19) \qquad \frac{\int_{o}^{\tau} f(T^{t}P)dt}{\int_{o}^{\tau} g(T^{t}P)dt} \quad , \quad \frac{\int_{o}^{\tau} f(T^{t}P')dt}{\int_{o}^{\tau} g(T^{t}P')dt}$$

have the same limits as $\tau \to \infty$ if the two streamlines occurring here are pos. asymptotic. Write briefly f_t, g_t, f'_t, g'_t for the four
integrands, respectively. Then by virtue of lemma 2 and in consequence of (17) and (18), the quotients

$$a_t \; = \; \frac{f'_t - f_t}{g'_t} \quad , \quad b_t \; = \; \frac{g'_t - g_t}{g'_t}$$

tend to zero as t $\to \infty$. The difference of the two quotients (19)
may be written as

106

E.Hopf

$$\frac{\int_0^\tau a_t g_t' \, dt}{\int_0^\tau g_t' \, dt} \; - \; \frac{\int_0^\tau f_t \, dt}{\int_0^\tau g_t \, dt} \; \cdot \; \frac{\int_0^\tau b_t g_t' \, dt}{\int_0^\tau g_\tau' \, d\tau} \quad .$$

This equality and (12) show that the difference tends to zero as $\tau \to \infty$ [1]. Obviously, the same is true as $\tau \to -\infty$ if the two stream-lines are neg. asymptotic. Consequently, the two invariant sets

$$B_- = [P \mid f^{**}(P) \geq c] \quad , \qquad B_+ = [P \mid f^*(P) \geq c]$$

satisfy hypothesis b) of the principal lemma no matter what the value of the constant c is. (16) implies the validity of hypothesis c). From that lemma we can, therefore, infer that either B_+ or its complement has measure m = 0 . This holds whatever value c has or, in other words, f^* is constant a.e., the first theorem is thereby completely proved.

Lemma 3 permits again to state the theorem with a sharper conclusion but a stronger hypothesis: If γ is of first class, if f(P), g(P) > 0 are in L_m and if these functions have the continuity properties (17) and (18) then $\lim q_\tau(P) = \int f dm / \int g dm$ holds in every point $p \in \gamma$ for almost all directions (P) in p.

2e. ADDITIONAL RESULTS AND PROBLEMS.

The previous results have straightforward generalizations to complete n-dimensional manifolds $\mathcal{M}$ of scalar curvature minus one. The universal covering space is the interior of the unit sphere

$$x_1^2 + x_2^2 + \cdots + x_n^2 < 1$$

---

[1] This conclusion is right provided that the first factor $\int_0^\tau f_t \, dt / \int_0^\tau g_t \, dt$ of the second term remains bounded as $\tau \to \infty$. This

./.

E.Hopf

endowed with the metric

$$ds^2 = \frac{4\Sigma dx_i^2}{(1 - \Sigma x_i^2)^2} \, , \quad dV = \frac{2^n \prod dx_i}{(1 - \Sigma x_i^2)^n} \, .$$

Its isometries are the n-dimensional Moebius transformations that map that covering space onto itself. In the (2n-1)-dimensional space of the directed line-elements a measure m is defined by

$$dm = d\omega dV$$

where $d\omega$ is the element of volume on the sphere of directions from dV. It is invariant both under the induced isometries and under the geodetic flow. In our memoir [11] the reader finds the preceding theory completely developed in n dimensions. He also finds there the following theorem and its proof :

If $\mathcal{M}$ is an n-dimensional manifold as stipulated above and if $\mathcal{M}$ has finite volume v($\mathcal{M}$) then the geodetic flow on $\mathcal{M}$ is strongly mixing: For two arbitrary sets A, B. of finite m in the line-element space Ω of $\mathcal{M}$ there holds

$$\lim_{t \to \infty} m(B \cdot T^t A) = \frac{m(A)m(B)}{m(\Omega)}$$

It is natural to conjecture that, in the more general case where $\mathcal{M}$ is of first class but not necessarily of finite volume the geodetic flow is still strongly mixing in a properly generalized sense. If $m(\Omega) = \infty$ the preceding relation is presumably still valid but uninteresting. Is this relation still valid if the expressions on both sides are replaced by quotients of them using four sets A, B; A', B'?

---

is certainly true for almost all P. However, the interference of one more nulset can be avoided if we argue as follows. We can restrict the functions f still further by requiring that f/g be bounded. The factor in question is then bounded by the same bound for any P. Moreover, this additional restriction on f does not invalidate the statement that the f's are dense in L_m.

108

E.Hopf

Let us return to n = 2 dimensions. Our surfaces γ furnish a large variety of interesting flows but they are restricted by the constancy of their curvature. In our paper [11] it is shown that exactly the same ergodic theory remains valid if γ is a complete two-dimensional Riemannian manifold of variable negative curvature (γ is assumed to be of differentiability class C'''). There is the same basic dichotomy into two classes, and both first and second theorem remain literally valid provided that the curvature remains between negative bounds. The mixture property has, however, not yet been proved for variable curvature.

We would like to return once more to our surfaces γ of constant negative curvature. We mentioned above that the purely topological subdivision into two kinds is different from our subdivision into two classes. The first kind is actually a larger totality than the first class. Take a surface γ of first kind but of second class. Paul Koebe had already proved, in his famous memoirs on non-euclidean space-forms, that on any γ of first kind there exists a quasi-ergodic geodesic (streamline dense in Ω).[1] For a general flow that is ergodic ($f^* = $ const. for every f) the same thing is true; even "most" streamlines are quasi-ergodic. What

---

[1] Koebe's topological analysis of the geodetical flow on surfaces γ has been carried on, and several of his results sharpened, by Gottschalk and Hedlund [5]. There is the following result on "topological" mixture on surfaces γ of the first "kind" (this is sherper than Koebe's result mentioned above): If A is a non-void open subset of Ω then $T^t A$ becomes denser and denser in Ω as $t \to \infty$.

To Hedlund [8] the ergodic theory of the geodesics on surfaces γ owes the first (and an ingeneous proof it is) proof of the measure-theoretical mixture of the flow on a surface γ of finite area. The author's proof of the same fact for n-dimensional manifolds $\mathcal{M}$ (quoted above) would not have been possible without Hedlund's beautiful idea. The author's first proof of ergodic mixture for surfaces γ of finite area [13] had yielded mixture only in a less stringent sense.

E.Hopf

about the converse question: Does the existence of a quasiergodic
streamline imply ergodicity of the flow? The answer is "no". On a
surface γ of first kind but second class the geodetic flow is dissipative.

E.Hopf

R E F E R E N C E S

1.   C.CARATHEODORY: Bemerkungen zum Ergodensatz von G.Birkhoff.
      Sitzungsber. Bayr. Akad. Wiss., Math.-Naturwiss. Abt. 1944(1947)
      p.189-208.

2.   R.V.CHACON and D.S.ORNSTEIN: A general ergodic theorem. Illi-
      nois J. of Math. 4 (1960), p.153-160.

3.   J.L.DOOB: Asymptotic properties of Markoff transition probabi-
      lities. Trans. Amer. Math. Soc. 63 (1948), p.393-421

4.   Y.N.DOWKER: A new proof of the general ergodic theorem. Acta
      Sci. Math: Szeged 12 (1950) p.162-166.

5.   W.H.GOTTSCHALK and G.A.HEDLUND: Topological dynamics. Amer.
      Math. Soc. Colloquium Publications 36 (1955).

6.   P.R.HALMOS: An ergodic theorem. Proc. Nat. Acad. Sci. 32 (1946)
      p.156-161.

7.   P.R.HALMOS: Measure theory. New York 1958.

8.   G.A.HEDLUND: The dynamics of geodesic flows. Bull. Amer. Math.
      Soc. 45 (1939), p.241-260.

9.   E.HOPF: Zwei Sätze uber den Verlauf der Bewegungen dynamischer
      Systeme. Math. Annalen 103 (1930), p.710-719.

10.   E.HOPF: Ergodentheorie. Berlin 1937.

11.   E.HOPF: Statistik der geodätischen Linien in Mannigfaltigkei-
      ten negativer Krümmung. Berichte Sächs. Akad. Wiss., Math.-
      Naturwiss. Klasse 91 (1939), p.261-304.

12.   E.HOPF: Statistik der Lösungen geodätischer Probleme vom uns-
      tabilen Typus II. Math. Annalen 117 (1940), p.590-608.

13.   E.HOPF: Beweis der Mischungscharakters der geodätischen Strö-
      mung auf vollständigen Flächen der Krümmung minus Eins und en-
      dlicher Oberfläche. Sitzungsber. Preuss. Akad. Wiss., Phys.-
      math. Klasse 30 (1938).

14.   E.HOPF: The general temporally discrete Markoff process. J. Rat
      Mech. Anal. 3 (1954), p.13-45.

15.   E.HOPF: On the ergodic theorem for positive linear operators.
      J. reine und angew. Math. 205 (1960), p.101-106.

16.   W.HUREWICZ: Ergodic theorem without invariant measure. Ann. Math
      45 (1944), p.192-206.

17.   S.KAKUTANI: Ergodic theorems and the Markoff process with a
      stable distribution. Proc. Imper. Acad. Tokyo 16 (1940),
      p.49-54.

E.Hopf

18. S.KAKUTANI: Ergodic theory. Proc. Int. Congress Math. Cambridge, Mass., Vol.2 (1952), p.128-142.

19. J.C.OXTOBY: On the ergodic theorem of Hurewicz. Ann. Math. 49 (1948), p.872-884.

20. K.YOSHIDA and S.KAKUTANI: Operator-theoretical treatment of Markoff's process and mean ergodic theorem. Ann. Math. 42 (1941), p.188-228.

CENTRO INTERNAZIONALE MATEMATICO ESTIVO

(C.I.M.E.)

J O S E' M A S S E R A

LES EQUATIONS DIFFERENTIELLES LINEAIRES

DANS LES ESPACES DE BANACH

ROMA - Istituto Matematico dell'Università - 1960

J.Massera

INTRODUCTION

Nous nous proposons de faire un exposé sommaire du contenu de plusieurs Mémoires écrits en collaboration avec J.J.Sch\"affer et qui ont été publiés sous le titre général *Linear differential equations and functional analysis,* Parties I, II, etc. [2, 3, 4, 5, 6, 7, 10, 12]; les théorèmes, formules, etc. de ces Mémoires seront cités par la suite de la façon suivante: Théorème II.5.4, formule I.(2.1), etc. Il nous sera aussi nécéssaire de resumer les principaux résul- tats de J.J.Sch\"affer dans son Mémoire *Function spaces with transla- tions* [11], que nous citerons F.3.2,F.(4.1), etc.

L'objet d'étude sont les équations

$$\dot{x} + A(t)x = 0 \tag{1}$$

$$\dot{x} + A(t)x = f(t) \tag{2}$$

$$\dot{x} + A(t)x = h(x,t) \tag{3}$$

où $t \in J = [0,\infty)$; x, f, $h \in X$, un espace de Banach quelconque; $A(t)$, pour chaque t fixe, est un endomorphisme (continu) de X ; $A(t)$, $f(t)$ sont des fonctions intégrables (Bochner, au sens conve- nable) dans chaque sous-intervalle borné $J' \subset J$.

Dans les années 1930-35, O.Perron [8], K.P.Persidskii [9] et I.G.Malkin [1] établirent l'équivalence des propriétés suivantes (dans le cas dim $X < \infty$, $A(t)$ continue):

(P_1) Pour chaque f continue bornée dans J (l'espace formé par ces functions, avec la norme du suprémum, sera désigné dans ce qui suit par $\mathcal{C}$), toutes les solutions de (2) $\in \mathcal{C}$ (nous les appe- lerons des $\mathcal{C}$ -solutions).

(P_2) Pour chaque h continue, $\|h\| \leq \beta$, $\|h(x',t) - h(x'',t)\| \leq \lambda \|x' - x''\|$, avec β,λ suffisamment petites, toutes les solutions de (3) $\in \mathcal{C}$.

J.Massera

(P_3) Il existe une fonction de Lyapunov $V(x,t)$ définie positive, ayant une borne supérieure infiniment petite (c.à.d., il existe des fonctions continues positives $a(r)$, $b(r)$, nulles pour $r = 0$, telles que $a(\| x \|) \leq V(x,t) \leq b(\| x \|)$), telle que la dérivée $V'(x,t)$ de V le long des solutions de (1) est définie négative.

(P_4) Il existe des constantes positives N, ν telles que, pour toute solution $x(t)$ de (1) et tout couple $t \geq t_0 \geq 0$, on a

$$\| x(t) \| \leq N e^{-\nu(t-t_0)} \| x(t_0) \| \ .$$

(P_5) La solution $x = 0$ de (1) est uniformément asymptotiquement stable, c.à.d., il existent une constante A et une fonction positive $T(\epsilon)$ telles que, pour toute solution $x(t)$ de (1) on a $\| x(t) \| \leq A \| x(t_0) \|$ pour $t \geq t_0 \geq 0$, et $\| x(t) \| \leq \epsilon \| x(t_0) \|$ pour $t_0 \geq 0$, $t \geq t_0 + T(\epsilon)$.

En réalité, Perron a démontré l'équivalence des propriétés suivantes, plus générale que l'équivalence de (P_1) et (P_2) :

(Q_1) Pour chaque $f \in \mathcal{C}$ il y a une famille à k paramètres $(0 \leq k \leq \dim X)$ de $\mathcal{C}$-solutions de (2).

(Q_2) Sous les hypothèses de (P_2) il y a une famille à k paramètres de $\mathcal{C}$-solutions de (3).

Dans nos travaux I-III nous nous avons posé la question de formuler adéquatement des propriétés analogues à (P_3), (P_4), (P_5) de façon a rétablir l'équivalence avec (Q_1) , (Q_2) . Deuxièmement, nous avons généralisé ces propriétés au cas où $\dim X = \infty$ et l'hypothèse de continuité est remplacée par des hypothéses du type de Carathéodory. Troisièmement, nous avons fait quelques progrès dans le sens de remplacer l'hypothèse "pour chaque $f \in \mathcal{C}$ " dans (P_1) et (Q_1) par "pour chaque f appartenant à certains espaces fonctionnels (tels que $\mathcal{L}^p$, etc.)". Finalement, nous avons envisa gé plusieurs applications des théorèmes obtenus aux cas périodique

J.Massera

presque-périodique, etc.

Plus tard, dans les travaux F et IV, nous avons considéré une catégorie générale d'espaces fonctionnels qui constituent le contexte naturel des problèmes qui nous occupent; pour chaque couple de tels espaces $\mathcal{B}$, $\mathcal{D}$, nous avons énoncé la propriété (Q_1) de la façon générale : "pour chaque $f \in \mathcal{B}$ il y a au moins une solution de (2) appartenant à $\mathcal{D}$ (une $\mathcal{D}$ -solution)", et étudié l'équivalence de cette propriété avec les autres énoncées plus haut. Nous avons aussi éliminé plusieurs restrictions superflues et envisagé de nouveaux types de comportement asymptotique analogues à (P_4) (stabilité en moyenne et par tranches). Il y a encore un vaste plan d'études complémentaires dont quelques résultats ont été déjà publiés [6, 7, 10, 12].

Nous voulons souligner que la généralisation des résultats obtenus au cas dim $X = \infty$ et aux $A(t)$ discontinues n'est pas du tout l'aspect le plus important de nos travaux : presque tous les théorèmes gardent leur intérêt dans le cas classique dim $X < \infty$. Le passage à la dimension infinie est banal dans plusieurs cas; quelquefois il implique, au contraire, des difficultés techniques plus ou moins grandes. Pour un petit nombre de résultats l'hypothèse de dimension finie est essentielle.

Nous appelons $U(t)$ la solution-opérateur (endomorphisme) de l'équation

$$\dot{U} + A(t)U = 0$$

qui satisfait $U(0) = I$ (identité); il est facile de voir que $U(t)$ est inversible pour chaque $t \in J$, La solution de (1) par (x_o, t_o) est alors $x(t) = U(t)U^{-1}(t_o)x_o$. La solution de (2) est donnée par la méthode de la variation des constantes :

J.Massera

$$(\mathrm{I}.(2.5)) \qquad x(t) = U(t)[U^{-1}(t_0)x_0 + \int_{t_0}^{t} U^{-1}(\tau)f(\tau)d\tau]. \qquad (4)$$

Des calculs élémentaires (Lemme de Gronwall, etc.) conduisent aux inégalités

$$(\mathrm{IV}.(2.1)) \quad \|x(t)\| \le (\| x(\tau) \| + \int_{J'} \|f(\tau)\| \, d\tau).\exp(\int_{J'} \|A(\tau)\|d\tau) \quad (5)$$

$$(\mathrm{IV}.(2.2)) \quad \|x(t)\| \le (1^{-1}\int_{J'} \|x(\tau)\|d\tau + \int_{J'} \|f(\tau)\| \, d\tau).$$

$$\exp(\int_{J'} \|A(\tau)\|d\tau) , \qquad (6)$$

où $x(t)$ est une solution de (2), J' est un sous-intervalle de J de longueur 1 et $t,\tau \in J'$.

Dans ce qui suit nous utiliserons la notion de *distance angulaire* de deux vecteurs non-nuls de l'espace X , définie par
$$\gamma(x,y) = \|x\|x\|^{-1} - y\|y\|^{-1}\| .$$

ESPACES FONCTIONNELS

Nous désignerons par $\mathcal{L}$ l'espace des fonctions (plutôt : classes d'équivalence) définies dans J à valeurs réelles, mesurables et intégrables dans chaque sous-intervalle fini $J' \subset J$, avec la topologie de la convergence en moyenne dans chaque J' . C'est un espace vectoriel localement convexe de Fréchet (complet et métrisable).

Soit $\mathcal{V}$ un espace vectoriel localement convexe de Hausdorff et $\mathcal{U}$ un espace normé. Nous dirons que $\mathcal{U}$ est *plus fin* que $\mathcal{V}$ lorsque $\mathcal{U}$ est algébraiquement contenu dans $\mathcal{V}$ et la convergence dans la topologie de $\mathcal{U}$ implique la convergence dans la topologie de $\mathcal{V}$. Alors, pour chaque semi-norme π de $\mathcal{V}$ il existe une constante positive a_π telle que $\pi(y) \le a_\pi \|y\|_{\mathcal{U}}$, $y \in \mathcal{U}$. Si $\mathcal{V}$ lui-même est normé, il existe une constante a telle que $\|y\|_{\mathcal{V}} \le a\|y\|_{\mathcal{U}}$,

J.Massera

$y \in \mathcal{U}$; dans ce cas nous écrirons $\mathcal{U} \leq a \mathcal{V}$ (la boule unitaire $\Sigma(\mathcal{U})$ de $\mathcal{U}$ est contenue dans a fois $\Sigma(\mathcal{V})$).

Soit $f(t)$ une fonction définie dans J , $\tau > 0$. Les *translatées* (à droite, à gauche) de f , $T_\tau^+ f = g$, $T_\tau^- f = h$ sont définies par

$$g(t) = 0 \qquad , \qquad 0 \leq t < \tau \ ,$$
$$g(t) = f(t - \tau) \quad , \quad t \geq \tau \ ,$$
$$h(t) = f(t + \tau) \quad , \quad t \geq 0 \ .$$

Nous définissons maintenant les classes d'espaces fonctionnels que nous utiliserons par la suite : les *espaces fonctionnels admettant des translations* (cf. F). Nous dirons qu'un espace $\mathcal{F}$ de fonctions (plutôt : classes d'équivalence) définies dans J à valeurs réelles appartient à la classe T s'il est normé (notation pour la norme $|\ |_{\mathcal{F}}$) et si les conditions suivantes sont remplies :

a) $\mathcal{F}$ est plus fin que $\mathcal{L}$, c.à.d., pour chaque sous-intervalle fini $J' \subset J$ il existe un nombre positif $a(J')$ tel que $\int_{J'} |f(\tau)| d\tau \leq a(J') |f|_{\mathcal{F}}$, $f \in \mathcal{F}$;

b) $f \in \mathcal{F}$, $|g(t)| \leq |f(t)|$ presque partout implique $g \in \mathcal{F}$, $|g|_{\mathcal{F}} \leq |f|_{\mathcal{F}}$;

c) $\mathcal{F} \neq \{0\}$;

d) $f \in \mathcal{F}$, $\tau > 0$ implique $T_\tau^+ f \in \mathcal{F}$, $|T_\tau^+ f|_{\mathcal{F}} = |f|_{\mathcal{F}}$.

Si l'on a aussi

d$^\#$) $f \in \mathcal{F}$, $\tau > 0$ implique $T_\tau^- f \in \mathcal{F}$,

nous dirons que $\mathcal{F}$ appartient à la classe $T^\#$.

Nous pouvons étendre la définition de ces classes d'espaces au cas des fonctions à valeurs dans un espace de Banach quelconque. Pour une telle fonction f nous définissons la fonction norme $\|f\|$ par la relation $\|f\| (t) = \| f(t)\|$; c'est une fonction à valeurs

J.Massera

réelles. Alors, $\mathcal{F}$ étant un espace de classes T [$T^{\#}$], nous désignons par $\mathcal{F}(X)$ l'espace des fonctions f à valeurs dans X , fortement mesurables, telles que $\|f\| \in \mathcal{F}$. En général, les propriétés plus importantes de $\mathcal{F}$ s'étendent à $\mathcal{F}(X)$ (par exemple, si $\mathcal{F}$ est complet, $\mathcal{F}(X)$ l'est aussi).

Pour les applications aux équations différentielles (les solutions étant toujours continues) il semblerait naturel de considérer des espaces de fonctions continues ayant des propriétés analogues. On peut définir une telle classe T^{c} (cf. F) mais la condition d) implique la restriction peu naturelle f(0) = 0 . D'autre part, on peut montrer qu'il existe une relation étroite entre ces espaces et des espaces T correspondants (Théorème F.6.3), de telle façon qu'on peut remplacer les premiers par les seconds (par exemple, $\mathcal{C}$ par $\mathcal{L}^{\infty}$) .

Les espaces classiques $\mathcal{L}^{p}$, $1 \leq p \leq \infty$, appartiennent évidemment à la classe $T^{\#}$. D'autres exemples importants sont :

$\mathcal{M}$: espace des fonctions telles que $\sup \{\int_{t}^{t+1} |f(\tau)| d\tau : t \in J\} < \infty$, avec la norme du suprémum;

$\mathcal{L}_{o}^{\infty}$: sous-espace de $\mathcal{L}^{\infty}$ formé par les fonctions dont la limite essentielle à l'infini est nulle;

$\mathcal{M}_{o}$: sous-espace de $\mathcal{M}$ formé par les fonctions telles que $\lim_{t \to \infty} \int_{t}^{t+1} |f(\tau)| d\tau = 0$. Tous ces espaces sont complets.

La relation d'ordre "plus fin" définit dans T [$T^{\#}$] une structure de treillis dont nous désignerons les opérations par $\wedge$, $\vee$. La $\wedge$ d'une famille d'espaces est formée par les fonctions appartenant à l'intersection pour lesquelles le suprémum des normes est fini; on prend ce suprémum comme norme dans $\wedge$. La $\vee$ d'une famille d'espaces est la somme de la famille, la norme d'une fonction étant l'infimum de la somme des normes pour toutes les décompo-

J.Massera

sitions possibles de la fonction donnée dans les espaces de la famille. On peut démontrer (Théorème F.4.2) que la $\wedge$ d'une famille d'espaces de classe T [$T^\#$] existe toujours et ou bien elle est aussi de classe T [$T^\#$] ou bien $= \{0\}$, la dernière possibilité étant exclue pour les familles finies; la $\vee$ d'une famille d'espaces de classe $T^\#$ appartient aussi à $T^\#$, si elle existe; si les espaces sont de classe T et la famille est totalement ordonnée, la $\vee$ appartient à T , si elle existe. La $\wedge$ [$\vee$] de deux espaces complets est aussi complète (pour $\wedge$ c'est aussi vrai même si la famille est infinie).

Une notion importante est celle de *fermeture locale*. Soit $\mathcal{F}$ un espace de classe T [$T^\#$]; la boule $\Sigma(\mathcal{F})$, considérée comme ensemble de $\mathcal{L}$, est convexe, balancée (si $f \in \Sigma(\mathcal{F})$, $\lambda f \in \Sigma(\mathcal{F})$, $|\lambda| = 1$), bornée, radialement fermée (l'intersection de $\Sigma(\mathcal{F})$ avec une demi-droite $\{\lambda f : f \in \mathcal{L} , \lambda \geq 0\}$ est fermée). La fermeture de $\Sigma(\mathcal{F})$ (dans $\mathcal{L}$) ayant les mêmes propriétés, on peut la considérer comme la boule d'un nouvel espace lc$\mathcal{F}$, que nous appelerons la fermeture locale de $\mathcal{F}$. Il est facile de voir, par exemple, que lc $\mathcal{L}_0^\infty = \mathcal{L}^\infty$, lc$\mathcal{M}_0 = \mathcal{M}$. On a $\mathcal{F} \leq$ lc$\mathcal{F}$, lc lc$\mathcal{F}$ = lc$\mathcal{F}$. Un espace pour lequel $\mathcal{F} =$ lc$\mathcal{F}$ sera dit localement fermé; tel est le cas des espaces $\mathcal{L}^p$ et $\mathcal{M}$. On peut montrer (Théorèmes F.2.2 et F.4.4 et Corollaire F.4.1) que tout espace localement fermé est complet; que si $\mathcal{F} \in T$ [$T^\#$] , lc$\mathcal{F} \in T$ [$T^\#$] ; que les $\wedge$ et $\vee$ de familles finies d'espaces localement fermés le sont aussi.

On peut démontrer que parmi les espaces de classe T [$T^\#$] il y en a un, à savoir $\mathcal{M}$, qui est le moins fin de tous. Quant aux espaces les plus fins, la situation est plus compliquée : il n'y a pas d'espace le plus fin dans T [$T^\#$] [1] mais la sous-

____
(1)   Ceci a été démontré après la réunion de Varenna.

J.Massera

classe des espaces localement fermés dans $T^\#$ a un espace qui est le plus fin de tous (Théorème F.4.19). Cet espace, que nous désignerons par ζ , est celui des fonctions f telles que

$$\sum_{n=0}^{\infty} \text{sup ess } \{|f(t)|: n \leq t \leq n+1\} < \infty \; ; \; \text{la norme de } \zeta \text{ est équi-}$$

valente à la somme de la série (la somme elle-même ne satisfait pas la condition d)).

Soit $\mathcal{F} \in T^\#$ quelconque et $\mathcal{F}'$ l'ensemble des fonctions f mesurables telles que $\sup \{|\int_0^{\infty} fg \, dt| : g \in \Sigma(\mathcal{F})\} < \infty$; on peut vérifier que $\mathcal{F}'$ avec ce suprémum comme norme, est un espace de classe $T^\#$ localement fermé (Théorèmes F.4.16 et F.4.17). $\mathcal{F}'$ s'appelle l'*espace associé* à $\mathcal{F}$ (ou dual de Köthe). On a (Corollaire F.4.7) $\mathcal{F}'' = \text{lc } \mathcal{F}$; l'association est un anti-automorphisme involutoire du treillis des espaces localement fermés de classe $T^\#$, $\mathcal{M}' = \zeta$, $\zeta' = \mathcal{M}$ (Lemmes F.4.9 et F.4.10 et Théorèmes F.4.17 et F.4.18); on a aussi (Lemme F.4.11) $(\mathcal{L}^1)' = \mathcal{L}^{\infty}$, $(\mathcal{L}^{\infty})' = \mathcal{L}^1$.

Finalement, on peut voir que les *espaces d'Orlicz* sont localement fermés de classe $T^\#$ (Théorème F.5.1), que l'associé d'un espace d'Orlicz est équivalent en norme à l'espace d'Orlicz défini par la fonction de Young conjuguée (Lemme F.5.3) et que parmi les espaces d'Orlicz $\mathcal{L}^1 \wedge \mathcal{L}^{\infty}$ et $\mathcal{L}^1 \vee \mathcal{L}^{\infty}$ sont, respectivement, le plus fin et le moins fin.

ADMISSIBILITE'

Nous dirons que le couple $(\mathcal{B}, \mathcal{D})$ (on devrait plutôt écrire $(\mathcal{B}(X), \mathcal{D}(X))$) d'espaces fonctionnels est *admissible* pour l'équation (2) si pour chaque $f \in \mathcal{B}(X)$ il y a au moins une solution de (2), $x \in \mathcal{D}(X)$ (une $\mathcal{D}$-solution). $\mathcal{B}$ et $\mathcal{D}$ seront toujours par la

J.Massera

suite des espaces complets de classe $\mathcal{T}$, quoique quelques résultats sont aussi valables dans des conditions plus générales.

Nous dirons que le couple $(\mathcal{B}_1, \mathcal{D}_1)$ est *plus fort* [faible]
que le couple $(\mathcal{B}_2, \mathcal{D}_2)$ si $\mathcal{B}_1$ est moins [plus] fin que $\mathcal{B}_2$ et
$\mathcal{D}_1$ est plus [moins] fin que $\mathcal{D}_2$. Les espaces étant complets, ce-
ci équivaut à $\mathcal{B}_1 \supset \mathcal{B}_2$, $\mathcal{D}_1 \subset \mathcal{D}_2$. L'admissibilité du couple
$(\mathcal{B}_1, \mathcal{D}_1)$ est alors une hypothèse plus forte sur l'équation que
l'admissibilité de $(\mathcal{B}_2, \mathcal{D}_2)$.

Soit $X_{o\mathcal{D}}$ l'ensemble des valeurs initiales $x(0)$ des $\mathcal{D}$-solutions
$x(t)$ de l'équation homogène (1); c'est une variété linéaire dans X
et si, pour une f donnée, l'équation (2) admet une $\mathcal{D}$-solution
$x(t)$, l'ensemble des valeurs initiales de toutes ses $\mathcal{D}$-solutions
est $x(0) + X_{o\mathcal{D}}$. $X_{o\mathcal{D}}$ n'est pas en général un sous-espace (c.à.d.,
fermé); I.4.1 est un exemple de cette circonstance pour $\mathcal{D} = \mathcal{C}$.
Dans la suite nous ferons toujours l'hypothèse que $X_{o\mathcal{D}}$ est fermé
et qu'il admet un sous-espace complémentaire $X_{1\mathcal{D}}$. En réalité, on
peut se passer de cette dernière hypothèse, mais au prix de comple-
xités que nous voulons éviter dans cet exposé.

Dans ces conditions, si $(\mathcal{B}, \mathcal{D})$ est admissible, pour chaque
$f \in \mathcal{B}$ il y a une et une seule $\mathcal{D}$-solution $x(t)$ de (2) avec
$x(0) \in X_{1\mathcal{D}}$ et la correspondance $\mathcal{B} \to \mathcal{D}$ ainsi définie est évidemment linéaire. Un résultat fondamental est qu'elle est aussi bor-
née :

THEOREME 1 (Corollaire IV.2.2). *Si $(\mathcal{B}, \mathcal{D})$ est admissible, il existe une constante K > 0 telle que pour chaque $f \in \mathcal{B}$ il y a une et u-
ne seule solution $x(t)$ de (2) satisfaisant $x \in \mathcal{D}$, $x(0) \in X_{1\mathcal{D}}$, et
on a $\|x\|_{\mathcal{D}} \le K\|f\|_{\mathcal{B}}$.*

La démonstration de ce résultat repose sur le théorème de gra-
phe fermé suivant qui peut se déduire des inégalités (5) et (6) :

J.Massera

THEOREME 2 (Lemme IV.2.1). *Soit $\{f_n\} \subset \mathcal{L}(X)$ et x_n une solution de $\dot{x} + A(t)x = f_n$. Si $\{f_n\}$ et $\{x_n\}$ convergent dans $\mathcal{L}$ respectivement vers f , x , on a $\dot{x} + A(t)x = f$ et $x_n \to x$ uniformément dans chaque sous-intervalle borné.*

DICHOTOMIES

La généralisation naturelle de la propriété (P_4) est le type de comportement des solutions de (1) que nous avons appelé dichotomies (simples ou exponentielles) et que, comme nous verrons ensuite, est étroitement lié à l'admissibilité de certains couples pour l'équation (2). En vue de (P_5) on peut aussi désigner ces types de comportement par *stabilité conditionnelle uniforme* (simple ou asymptotique) de la solution $x = 0$ de (1); le cas asymptotique est essentiellement équivalent, pour les équations linéaires, a ce que N.N.Krasovskii appelle *comportement non-critique uniforme* (cf. Théorème I.3.5, Corollaire I.3.1 et Exemples I.3.3 et I.3.4).

Nous dirons que les sous-espaces complémentaires Y_0, Y_1 de X [1] induisent une *dichotomie exponentielle* des solutions de (1) s'il existe des constantes positives N , N' , ν , ν' , γ_0 telles que les conditions suivantes sont remplies :

(Ei) Pour chaque solution $x(t)$ de (1) avec $x(0) \in Y_0$ on a
$$\|x(t)\| \leq N e^{-\nu(t-t_0)} \|x(t_0)\| , \ t \geq t_0 \geq 0 ;$$

(Eii) Pour chaque solution $x(t)$ de (1) avec $x(0) \in Y_1$ on a
$$\|x(t)\| \geq N' e^{\nu'(t-t_0)} \|x(t_0)\| , \ t \geq t_0 \geq 0 ;$$

(Eiii) Pour chaque couple de solutions $x_i(t)$ de (1) avec $x_i(0) \in Y_i$, $i = 0, 1$, on a $\gamma(x_0(t), x_1(t)) \geq \gamma_0$, $t \geq 0$.

Si dans (Ei), (Eii) on supprime les facteurs exponentiels, on obtient les propriétés (Di), (Dii), lesquelles, avec (Diii) = (Eiii),

---
[1]
On peut formuler la définition de façon a ne pas faire intervenir Y_1

J.Massera

définissent le comportement appelé *dichotomie (simple)* des solutions de (1) induite par les sous-espaces Y_0, Y_1 .

On peut démontrer (Lemme I.5.4) que si $A \in M$ alors (Ei) et (Eii) impliquent (Eiii). Dans le cas le plus simple, on a :

LEMME 1. *Si* dim $X < \infty$ *, A = const. et si les parties réelles des racines caractéristiques de* A *sont non-nulles, on a une dichotomie exponentielle; et réciproquement. Si aux racines caractéristiques purement imaginaires correspondent des diviseurs élémentaires li-néaires, alors on a une (et peut-être plusieurs) dichotomie simple; et réciproquement.*

On peut aussi envisager des comportements "en moyenne" ou "par tranches" du type suivant :

On a la *stabilité asymptotique uniforme en moyenne* s'il exis-tent des sous-espaces complémentaires Y_0 , Y_1 , des constantes po-sitives ν , ν' et des fonctions positives $M(\Delta)$, $M'(\Delta)$ ne dépendent que de $\Delta > 0$, tels que :

(Mi) Pour chaque solution $x(t)$ de (1) avec $x(0) \in Y_0$ on a

$$\int_t^{t+\Delta} \|x(u)\| \, du \leq M(\Delta) \, e^{-\nu(t-t_0)} \int_{t_0}^{t_0+\Delta} \|x(u)\| \, du , \quad t \geq t_0 \geq 0 ;$$

(Mii) Pour chaque solution $x(t)$ de (1) avec $x(0) \in Y_i$ on a

$$\int_t^{t+\Delta} \|x(u)\| \, du \geq M'(\Delta) e^{\nu'(t-t_0)} \int_{t_0}^{t_0+\Delta} \|x(u)\| \, du , \quad t \geq t_0 \geq 0 .$$

On a la *stabilité asymptotique uniforme par tranches* (relati-vement à un espace fonctionnel donné $\mathcal{D}$ contenant les fonctions caractéristiques des sous-intervalles bornés) s'il existent Y_0 , Y_1 , ν , ν' , $M(\Delta)$, $M'(\Delta)$ comme plus haut tels que :

---

qui peut même ne pas exister comme complément de Y_0 ; le fait que Y_0 est fermé est, par contre, essentiel.

J.Massera

(Ti) Pour chaque solution $x(t)$ de (1) avec $x(0) \in Y_o$ on a

$$\left\| \chi_{[t,t+\Delta]} x \right\|_{\mathcal{D}} \leq M(\Delta) e^{-\nu(t-t_o)} \left\| \chi_{[t_o,t_o+\Delta]} x \right\|_{\mathcal{D}} \quad , \quad t \geq t_o \geq 0 \quad ,$$

(où χ_E désigne la fonction caractéristique de l'ensemble $E \subset J$);

(Tii) Pour chaque solution $x(t)$ de (1) avec $x(0) \in Y_1$ on a

$$\left\| \chi_{[t,t+\Delta]} x \right\|_{\mathcal{D}} \geq M'(\Delta) e^{\nu'(t-t_o)} \left\| \chi_{[t_o,t_o+\Delta]} x \right\|_{\mathcal{D}} \quad , \quad t \geq t_o \geq 0 \quad .$$

Si l'on supprime les facteurs exponentiels, on obtient les comportements *simples* respectifs. Il n'y a pas apparemment de correspondant naturel de la propriété (Eiii) dans les cas envisagés maintenant.

THEOREMES FONDAMENTAUX

On peut démontrer les théorèmes suivants :

THEOREME 3 (Théorèmes IV.6.1 et IV.6.2). *Si $(\mathcal{B},\mathcal{D})$ est un couple admissible d'espaces de classe T (un T-couple), il y a stabilité uniforme en moyenne et par tranches des solutions de (1) avec $Y_o = X_{o\mathcal{D}}$, $Y_1 = X_{1\mathcal{D}}$. Si $(\mathcal{B},\mathcal{D})$ n'est pas plus faible que $(\mathcal{L}^1 , \mathcal{L}^\infty_o)$ la stabilité est même asymptotique.*

THEOREME 4 (Théorème IV.8.1). *Si $(\mathcal{B},\mathcal{D})$ est un T-couple admissible plus fort que $(\mathcal{L}^1 , \mathcal{L}^\infty)$, $X_{o\mathcal{D}}$, $X_{1\mathcal{D}}$ induisent une dichotomie des solutions de (1).*

THEOREME 5 (Théorème IV.8.2). *Si $(\mathcal{B},\mathcal{D})$ est un T-couple admissible et $A \in \mathcal{M}$, $X_{o\mathcal{D}}$, $X_{1\mathcal{D}}$ induisent une dichotomie des solutions de (1).*

THEOREME 6 (Théorème IV.8.3). *S'il y a une dichotomie des solutions de (1), le couple $(\mathcal{L}^1 , \mathcal{L}^\infty)$ est admissible (et tous les couples plus faibles); si $\dim Y_o < \infty$, même le couple $(\mathcal{L}^1 , \mathcal{L}^\infty_o)$ est admissible.*

126

J.Massera

THEOREME 7 (Théorèmes IV.9.1 et IV.9.2). *Si* $(\mathcal{B}, \mathcal{D})$ *est un* T *-couple admissible qui n'est pas plus faible que* $(\mathcal{L}^1, \mathcal{L}_o^\infty)$ *et si ou bien il y a une dichotomie des solutions de* (1) *ou bien* $A \in \mathcal{M}$ *, alors* $X_{o\mathcal{D}}$ *,* $X_{1\mathcal{D}}$ *induisent une dichotomie exponentielle des solutions de* (1).

THEOREME 8 (Théorèmes V.5.2 et IV.9.3). *S'il y a une dichotomie exponentielle des solutions de* (1), *pour que le* T *-couple* $(\mathcal{B}, \mathcal{D})$ *soit admissible il suffit (et, si* $A \in \mathcal{M}$ *, il le faut) que* $f \in \mathcal{B}$ *implique* $F \in \mathcal{D}$ *où* $F(t) = \int_t^{t+1} |f(u)| du$ *. En particulier, les couples trés forts* $(\mathcal{M}, \mathcal{L}^\infty)$ *,* $(\mathcal{M}_o, \mathcal{L}_o^\infty)$ *,* $(\mathcal{L}^1, \mathcal{C})$ *sont admissibles.*

Il est impossible de donner ici même un apperçu synthétique des démonstrations de ces théorèmes, qui sont assez longues et compliquées. Nous voulons seulement indiquer l'idée maîtresse de la démonstration. Pour les théorèmes directs (4, 5 et 7), si $x(t)$ est une solution de l'équation (1) avec, disons, $x(0) \in X_{o\mathcal{D}}$ (c.à.d., une $\mathcal{D}$-solution) et si $\phi(t)$ est une fonction scalaire dont la dérivée est nulle en dehors d'un sous-intervalle borné et telle que $\phi(0) = 0$, la fonction $y(t) = \phi(t)x(t)$ est une solution de (2) correspondant à $f(t) = \dot{\phi}(t)x(t)$; avec un ϕ convenable on obtient $f \in \mathcal{B}$ et, puisque $y(0) = 0 \in X_{1\mathcal{D}}$, on peut appliquer le Théorème 1, on obtient $|\phi x|_{\mathcal{D}} \leq K |\phi x|_{\mathcal{B}}$ et les propriétés de type dichotomique s'ensuivent. Pour les Théorèmes réciproques (6 et 8), la base de la démonstration est la formule de la variation des constantes (4) écrite comme il suit :

$$x(t) = \int_o^t U(t)Q_o U^{-1}(\tau)f(\tau)d\tau - \int_t^\infty U(t)Q_1 U^{-1}(\tau)f(\tau)d\tau \ ,$$

où Q_o, Q_1 sont les projections complémentaires sur Y_o, Y_1 ; $U(t)Q_o U^{-1}(\tau)f(\tau)$ et $U(t)Q_1 U^{-1}(\tau)f(\tau)$ étant, pour chaque τ fixe, des solutions de l'équation homogène qui partent, respectivement

J.Massera

de Y_0 et Y_1 , on peut alors leur appliquer les inégalités caractéristiques des dichotomies.

AUTRES RESULTATS

Les théorèmes suivants peuvent donner une idée des applications de la méthode et de quelques compéments possibles.

a) Théorèmes sur l'existence de solutions presque-périodiques [2, 6] :

THEOREME 9 (Théorème I.6.2). *S'il y a une dichotomie exponentielle et si* A(t) *est presque-périodique, pour chaque f presque-périodique l'équation* (2) *admet une et une seule solution presque-périodique.*

b) Théorèmes sur les équations quasi-linéaires (cf. propriété (Q_2) et [2, 7]) :

THEOREME 10. *Soit* $(\mathcal{B},\mathcal{D})$ *un* $\mathcal{T}$*-couple admissible et* h(x,t) *une fonction définie pour* $t \in J$, $x \in X$, $\|x\| < a$ $(0 < a \leq \infty)$, *à valeurs dans* X , *telle que pour chaque fonction* $x \in \mathcal{D} \cap \mathcal{L}^{\infty}$, $|x|_{\infty} < a$ *on ait* $h(x(t),\ t) \in \mathcal{B}$. *Soit* $\beta = |h(0,t)|_{\mathcal{B}}$ *et admettons qu'il existe une constante positive* λ *telle que, pour chaque couple* x', $x'' \in \mathcal{D} \cap \mathcal{L}^{\infty}$, $|x'|_{\infty}$, $|x''|_{\infty} < a$, *on ait*

$$|h(x'(t),\ t) - h(x''(t),\ t)|_{\mathcal{B}} \leq \lambda |x' - x''|_{\mathcal{D}} .$$

Alors, si β, λ *sont suffisamment petites, il existe un* $b > 0$ *tel que pour chaque* $\xi_0 \in X_{o\mathcal{D}}$, $\|\xi_0\| < b$, *il y a une et une seule solution* $x(t,\ \xi_0) \in \mathcal{D}$ *de* (3) *telle que* $x(0,\ \xi_0) = \xi_0 + \xi_1$, $\xi_1 \in X_{1\mathcal{D}}$.

COROLLAIRE. *Soit* $A \in \mathcal{M}$ *et* $(\mathcal{B},\mathcal{D})$ *un* $\mathcal{T}$*-couple admissible qui n'es pas moins faible que* $(\mathcal{L}^1,\ \mathcal{L}^{\infty}_o)$. *Soit* h(x,t) *définie pour* $t \in J$, $x \in X$, $\|x\| < a$ $(0 < a \leq \infty)$; *admettons que pour chaque* x *fixe elle*

J.Massera

est une fonction de t continue et bornée et qu'il existe une constan-
te λ telle que $\|h(x',t) - h(x'',t)\| \leq \lambda \|x' - x''\|$ pour x', x'' $\in$ X ,
$\|x'\|$, $\|x''\| <$ a , t $\geq$ 0 . Alors, la conclusion du Théorème 10 reste
valable si l'on remplace $\mathcal{D}$ par $\mathcal{C}$.

THEOREME 11. *Soit X réflexif et supposons qu'il y a une dicho-*
tomie exponentielle des solutions de (1). Soit h(x,t) une fonction
faiblement continue de X $\times$ J dans X (c.à.d., continue lorsque X est
muni de la topologie faible); admettons que $\|h(x,t)\| \leq \phi(\|x\|)$, $\phi(r)$
étant une fonction continue définie pour r $\geq$ 0 , $\phi(r) = o(r)$ pour
r $\to$ ∞ . Alors, pour chaque $\xi_o \in Y_o$, l'equation (3) admet au moins
une solution bornée $\mathbf{x}(t, \xi_o)$ telle que $x(0, \xi_o) = \xi_o + \xi_1$, $\xi_1 \in Y_1$.

Des résultats analogues sont valables pour l'existence de solutions presque-périodiques d'équations quasi-linéaires presque-périodiques.

c) Grossièreté [2] :

THEOREME 12. *Soit ($\mathcal{B}, \mathcal{L}^{\infty}$) admissible et B $\in$ $\mathcal{B}(\tilde{X})$ ($\tilde{X}$: l'es-*
pace des endomorphismes de X). Alors, si $\|B\|_{\mathcal{B}}$ est suffisamment pe-
tite, ($\mathcal{B}, \mathcal{L}^{\infty}$) est admissible pour $\hat{x} + (A(t) + B(t))x = f(t)$.

On peut aussi voir (Théorème I.8.2) que l'ensemble $\Omega \in \mathcal{M}$ des
A(t) pour lesquels il y a dichotomie exponentielle est un ouvert et
que les éléments caractéristiques de la dichotomie (Y_o, Y_1, ν, ν',
etc.) possèdent des propriétés de continuité dans leur dépendence
par rapport à A $\in$ Ω .

d) Méthode de Lyapunov (cf. propriété (P_3) et [4]) :

THEOREME 13 (Théorèmes III.3.1 et III.3.2). *Soit A $\in$ $\mathcal{M}$. Con-*
dition nécéssaire et suffisante pour qu'il y ait une dichotomie ex-
ponentielle est qu'il existe une fonction de Lyapunov (généralisée)
V ayant une borne supérieure infiniment petite et telle que V' soit
définie négative.

J.Massera

THEOREME 14 (Théorèmes III.4.3 et III.4.4). *Soit* dim X < ω .
*Condition nécéssaire et suffisante pour qu'il y ait une dichotomie
est qu'il existe deux fonctions de Lyapunov non-négatives* V_o, V_1,
positivement homogènes en x du même degré, telles que $V_o + V_1$
*soit définie positive et possède une borne supérieure infiniment
petite et* $V_o' \leq 0$, $V_1' \geq 0$.

e) Cas où A(t) est pèriodique (résultats essentiellement nouveaux seulement lorsque dim X = ∞) [3, 7] :

THEOREME 15 (Théorème II.2.1). *Soit* A(t) *périodique de période*
1 *et* $|A|_m <$ log 4 ; *alors il existe une représentation de Floquet
des solutions de* (1) : $x(t) = P(t)e^{Bt}x_o$ *avec P périodique de période* 1 , B = *const.*

La représentation de Floquet n'est pas toujours possible si
dim X = ∞ et $|A|_m > \pi$ (Exemple II.2.1).

THEOREME 16 (Corollaire II.3.3).*Soit* A(t) *périodique de période*
1 *et supposons que la fermeture de* $X_{o\mathcal{C}}$ *soit réflexive. Etant donné
une f périodique de période* 1 , *si l'équation* (2) *admet une solution bornée, elle admet une solution périodique de période* 1 .

THEOREME 17 (Théorème II.3.5). *S'il existe un* T-*couple admis-
sible qui n'est pas plus faible que* ($\mathcal{L}^1$, $\mathcal{L}_o^\infty$) *et si* A(t), f(t) *sont
périodiques de période* 1 , *l'équation* (2) *admet une et une seule so-
lution périodique de période* 1.

f) Cas où A est constant (cf. Lemme 1 et [10]) :

THEOREME 18. *Si* A = *const., pour qu'il y ait dichotomie exponentielle des solutions de* (1) *il faut et il suffit que le spectre de
A ne coupe pas l'axe imaginaire.*

BIBLIOGRAPHIE

1. I.G.MALKIN, *On stability in the first approximation*, Sbornik Nauonyh Trudov Kazanskogo Aviacionnogo Instituta, 3 (1935), 7-17 (en russe).

2. J.L.MASSERA and J.J.SCHÄFFER, *Linear differential equations and functional analysis*, I, Annals of Math., 67 (1958), 517-573.

3. J.L.MASSERA and J.J.SCHÄFFER, *Linear differential equations and functional analysis*, II. *Equations with periodic coefficients*, ibid, 69 (1959), 88-104.

4. J.L.MASSERA and J.J.SCHÄFFER, *Linear differential equations and functional analysis*, III. *Lyapunov's second method in the case of condicional stability*, ibid, 69 (1959), 535-574.

5. J.L.MASSERA and J.J.SCHÄFFER, *Linear differential equations and functional analysis*, IV, Math.Annalen, 139 (1960), 287-342.

6. J.L.MASSERA, *Un criterio de existencia de soluciones casi-periodicas de ciertos sistemas de ecuaciones diferenciales casi-periodicas*, Publ.Inst.Mat.Estad. (Montevideo), III (1958), 99-103.

7 J.L.MASSERA, *Sur l'existence de solutions bornées et périodiques des systèmes quasi-linéaires d'équations différentielles*, Annali di Mat. (IV) 51 (1960), 95-106.

8. O.PERRON, *Die Stabilitätsfrage bei Differentialgleichungen*, Math. Zeitschrift, 32 (1930), 703-728.

9. K.P.PERSIDSKII, *On the stability of motion in the first approximation*, Mat.Sbornik, (N.S.) 40 (1933), 284-293 (en russe).

10. J.J.SCHÄFFER, *Ecuaciones diferenciales lineales con coeficientes constantes en espacios de Banach*, Publ. Inst. Mat. Estad. (Montevideo), III (1958), 105-110.

11. J.J.SCHÄFFER, *Function spaces with translations*, Math.Annalen, 137 (1959), 209-262; *Addendum*, ibid, 138 (1959), 141-144.

12. J.J.SCHÄFFER, *Linear Differential equations and functional analysis*, V, Math.Annalen, 140 (1960), 308-321.

CENTRO INTERNAZIONALE MATEMATICO ESTIVO

(C.I.M.E.)

LUIGI AMERIO

FUNZIONI QUASI-PERIODICHE ASTRATTE E PROBLEMI
DI PROPAGAZIONE

ROMA - Istituto Matematico dell'Università - 1960

133

FUNZIONI QUASI-PERIODICHE ASTRATTE E PROBLEMI

DI PROPAGAZIONE

di L.AMERIO

1.- Esporremo alcuni recenti risultati sulle soluzioni quasi-periodiche (q.p.) dell'equazione delle onde, e, più in generale, delle equazioni differenziali q.p. astratte.

Ricordiamo, innazi tutto, che *una funzione continua* :

$$(1.1) \qquad y = f(t) , \qquad t \in J = (- \overline{\infty + \infty}) ,$$

a valori in uno spazio B , di Banach, *è quasi-periodica se ad ogni $\epsilon > 0$ può farsi corrispondere un insieme T_ϵ, relativamente denso, di numeri τ (quasi-periodi) per ciascuno dei quali risulti*

$$(1.2) \qquad \underset{t \in J}{Sup} \ \| f(t + \tau) - f(t) \| \leq \epsilon .$$

Questa definizione si riduce, se B è euclideo, a quella, classica, di Bohr. Vale inoltre, anche nel caso astratto, il fondamentale *criterio* di Bochner [1] : *condizione necessaria e sufficiente perchè f(t), continua in J , sia q.p. è che da ogni successione reale $\{h_n\}$ possa estrarsi una sottosuccessione $\{h'_n\}$ tale che la successione $\{f(t + h'_n)\}$ converga uniformemente in J .*

Per quanto si dirà nei §§2 e 3, interessa supporre B *hilbertiano*: precisamente interessano gli spazi L^2, H_0^1, E di cui ora ricorderemo le definizioni.

I) L^2 è lo spazio delle funzioni reali $y = y(x)$ a quadrato integrabile in un insieme Ω (*aperto, limitato e connesso*) dello spazio euclideo X_m ($x = x_1, \ldots, x_m$), con la consueta definizione di prodotto scalare :

135

L.Amerio

$$(1.3) \qquad (y(x), z(x))_{L^2} = \int_\Omega y(x)\, z(x)\, d\Omega \;.$$

II) H_0^1 è lo spazio delle funzioni reali $y = y(x)$ a quadrato integrabile in Ω insieme alle derivate prime (nel senso della *teoria delle distribuzioni*) e nulle (in senso *generalizzato*) sulla frontiera σ di Ω .

Assumeremo come prodotto scalare, in H_0^1, la quantità

$$(1.4) \quad (y(x),\, z(x))_{H_0^1} = \int_\Omega \Big\{ \sum_{j,k}^{1\,-\,m} a_{jk}(x)\, \frac{\partial y}{\partial x_j}\, \frac{\partial z}{\partial x_k} + a(x)\,y(x)\,z(x) \Big\} d\Omega \;.$$

Nella (1.4) le $a_{jk}(x) = a_{kj}(x)$, $a(x)$ sono funzioni *misurabili e limitate*, è $a(x) \geq 0$ e *la forma quadratica* $\sum_{j,k}^{1\,-\,m} a_{jk}(x)\, \xi_j\, \xi_k$ *soddisfa alla limitazione*

$$\sum_{j,k}^{1\,-\,m} a_{jk}(x)\, \xi_j\, \xi_k \geq \mu \sum_1^{\overset{m}{}} \xi_j^2 \qquad (\mu > 0).$$

Ricordiamo che H_0^1 si ottiene *completando*, rispetto alla norma indotta dalla (1.4), lo spazio delle funzioni continue in Ω insieme alle loro derivate prime, e nulle in un intorno della frontiera σ . E' inoltre $H_0^1 \subset L^2$.

III) E è lo spazio prodotto cartesiano di H_0^1 per L^2 :

$$E = H_0^1 \times L^2 \;.$$

Ogni elemento $Y(x) \in E$ è perciò costituito da una *coppia* $\{y_0(x),\, y_1(x)\}$ di funzioni : $y_0(x) \in H_0^1$, $y_1(x) \in L^2$.

Si assumerà come prodotto scalare, in E, la quantità,

$$(1.5) \quad (Y(x),\, Z(x))_E = (y_0(x);\, z_0(x))_{H_0^1} + (y_1(x),\, z_1(x))_{L^2} \;,$$

cui corrisponde la norma

$$(1.6) \quad \|Y(x)\|_E = \{ \|y(x)\|_{H_0^1}^2 + \|y_1(x)\|_{L^2}^2 \}^{1/2} \;.$$

L.Amerio

Chiameremo E lo *spazio dell'energia* e la metrica (1.6) la *metrica dell'energia.*

Infatti le quantità

$$\frac{1}{2} \, \|y_0(x)\|_{H^1_0}^2 \quad , \quad \frac{1}{2} \, \|y_1(x)\|_{L^2}^2 \quad , \quad \frac{1}{2} \, \|Y(x)\|_E^2$$

rappresentano rispettivamente l'*energia potenziale, l'energia cinetica e l'energia totale* di una membrana Ω , col bordo fisso, in cui $y_0(x)$ designi lo *spostamento* del punto x, $y_1(x)$ la *velocità* del punto medesimo.

2.- Il problema misto, secondo Hadamard, per l'*equazione delle onde* (o *equazione della membrana vibrante*) :

$$(2.1) \qquad \frac{\partial^2 y}{\partial t^2} = \sum_{j,k}^{1-\cdots m} \frac{\partial}{\partial x_j} \left(a_{jk}(x) \frac{\partial y}{\partial x_k} \right) - a(x)y + f(x, t)$$

$$(t \in J, \ x \in \Omega)$$

è stato oggetto di numerose ricerche, anche recenti, [2]. Per la nostra esposizione interessa considerare le *soluzioni deboli* della (2.1) : ci riferiremo inoltre al problema definito dalle *condizioni iniziali*

$$(2.2) \qquad y(x,0) = y_0(x) \quad , \quad y_t(x,0) = y_1(x)$$

con $y_0(x)$, $y_1(x)$ funzioni assegnate) e dalla *condizione ai limiti*

$$(2.3) \qquad y(x,t)\big|_{\sigma} = 0 \ ,$$

corrispondente alla vibrazione di una *membrana col bordo fisso.* Le soluzioni deboli si associano alla teoria *variazionale,* anzichè *differenziale,* della membrana vibrante: sotto questo punto di vista, le soluzioni deboli ci appaiono come le *vere* soluzioni,

L.Amerio

risultando, in più, soluzioni della (2.1) (soluzioni da dirsi *forti*) quando soddisfino, unitamente alla frontiera σ , a convenienti condizioni di regolarità.

Soluzioni *deboli* della (2.1) saranno dette, precisamente, le soluzioni $y(x,t)$ dell'*equazione variazionale delle onde* (con la quale si traduce *il principio di* Hamilton):

$$(2.4) \quad \int_\alpha^\beta \{ (y_t(x,t), \phi_t(x,t))_{L^2} - (y(x,t), \phi(x,t))_{H_0^1} +$$
$$+ (f(x,t), \phi(x,t))_{L^2} \} \, dt = 0 .$$

Questa deve essere verificata (*comunque si prenda l'intervallo* $\alpha \longmapsto \beta$) in corrispondenza di *tutte le variazioni* $\phi(x,t)$, *nulle per* $t = \alpha$ *e per* $t = \beta$. In modo preciso, cercheremo le soluzioni del problema misto nella classe Γ delle funzioni $u(x,t)$ soddisfacenti alle seguenti condizioni:

A) $u(x,t)$ sia *continua,* come funzione di t a valori in H_0^1;

B) $u(x,t)$ sia *derivabile,* come funzione di t a valori in L^2, e la *derivata* $u_t(x,t)$ risulti *continua in* J .

E' chiaro che, *se valgono le* A) *e* B), *la funzione* $U(x,t) = \{ u(x,t), u_t(x,t) \}$, *a valori in* E , *ha ivi, come traiettoria, una linea continua.*

Sulla variazione $\phi(x,t)$ si fanno ancora, nell'intervallo $\alpha \longmapsto \beta$, le ipotesi A), B); si supporrà, in più, che sia

$$\| \phi(x,\alpha) \|_{H_0^1} = \| \phi(x,\beta) \|_{H_0^1} = 0 .$$

Quanto al termine noto $f(x,t)$, ammetteremo che sia a valori in L^2, per quasi tutti i t , e che *la norma* $\| f(x,t) \|_{L^2}$ *sia integrabile in ogni intervallo limitato.* Infine, per quanto riguarda i dati iniziali, si supporrà che, nelle (2.2), sia

L.Amerio

$$y_0(x) \in H_0^1, \qquad y_1(x) \in L^2,$$

cioè

$$Y(x,0) \in E.$$

Si può allora dimostrare (cfr. L.AMERIO, Loc. cit. in [2]) che *la soluzione* $Y(x,t) = \{y(x,t), y_t(x,t)\}$ *della* (2.4), *con la condizione iniziale*

$$Y(x,0) = \{y_0(x), y_1(x)\},$$

esiste in tutto J, *è unica e dipende con continuità dai dati,* in virtù della formula di maggiorazione

$$(2.5) \qquad \|Y(x,t)\|_E \leq \|Y(x,0)\|_E + \left| \int_0^t \|f(x,\eta)\|_{L^2} d\eta \right|.$$

La soluzione medesima può calcolarsi, seguendo lo schema classico delle soluzioni elementari, mediante la formula risolutiva :

$$(2.6) \qquad Y(x,t) = \left\{ \sum_1^\infty \gamma_n(t) \frac{u_n(x)}{\lambda_n}, \quad \sum_1^\infty \frac{\gamma_n'(t)}{\lambda_n} u_n(x) \right\}.$$

Nella (2.6) le $u_n(x)$ e le costanti λ_n rappresentano rispettivamente le *autosoluzioni* e gli *autovalori* dell'equazione

$$(2.7) \qquad (u(x), h(x))_{H_0^1} = \lambda^2 (u(x), h(x))_{L^2},$$

che deve intendersi soddisfatta per *tutte* le $h(x) \in H_0^1$.

La (2.7) ammette una successione $\{\lambda_n\}$ di autovalori, per i quali risulta

$$0 < \lambda_1 \leq \lambda_2 \leq \ldots \leq \lambda_n \leq \ldots, \quad \lim_{n\to\infty} \lambda_n = +\infty,$$

e le corrispondenti autosoluzioni sono legate dalle condizioni di ortogonalità

$$(u_r(x), u_s(x))_{L^2} = \left(\frac{u_r(x)}{\lambda_r}, \frac{u_s(x)}{\lambda_s} \right)_{H_0^1} = \delta_{rs}$$

L. Amerio

Inoltre la successione $\{u_n(x)\}$ è *completa* sia in L^2 che in H^1_0 .

Risulta infine

$$(2.8) \qquad \gamma_n(t) = a_n \cos \lambda_n t + b_n \sin \lambda_n t + \int_0^t f_n(\eta)\sin \lambda_n(t-\eta)d\eta,$$

essendo

$$(2.9) \qquad a_n = (y_0(x), \frac{u_n(x)}{\lambda_n})_{H^1_0} , \quad b_n = (y_1(x), u_n(x))_{L^2} ,$$

$$(2.10) \qquad f_n(t) = (f(x,t), u_n(x))_{L^2} .$$

Dalle (2.9) segue :

$$(2.11) \quad \sum_1^\infty \left(a_n^2 + b_n^2\right) = \|y_0(x)\|^2_{H^1_0} + \|y_1(x)\|^2_{L^2} = \|Y(x,0)\|^2_E < +\infty .$$

Ricordiamo infine che, *se* $Y(x,t)$ *e* $Z(x,t)$ *sono due soluzioni corrispondenti ai termini noti* $f(x,t)$, $g(x,t)$, *vale l'eguaglianza :*

$$(2.12) \quad \frac{d}{dt}(Y(x,t), Z(x,t))_E = (f(x,t), z_t(x,t))_{L^2} + (g(x,t), y_t(x,t))_{L^2}$$

La (2.12), per $f(x,t) = g(x,t) = 0$, $Y(x,t) = Z(x,t)$ diventa

$$(2.13) \qquad \frac{d}{dt} \|Y(x,t)\|^2_E = 0 ,$$

che esprime il *principio di conservazione dell'energia.*

3.- La *quasi-periodicità* degli integrali dell'equazione *omogenea* delle onde ($f(x,t) = 0$) è stata dimostrata, in ipotesi sempre più generali, da vari Autori: Muckenhoupt [3] (per m = 1), Bochner [4], Bochner e von Neumann [5], Sobolev [6], Ladyzenskaja [7].

Consideriamo ora l'equazione *non omogenea*, supponendo *il termine noto* $f(x,t)$ *funzione q.p. di* t , *a valori in* L^2 . Zaidman [8]

L.Amerio

ha dato, in questo caso, per il primo, una condizione *sufficiente*
perchè le soluzioni Y(x,t) siano q.p., come funzioni a valori in
E : valendosi, precisamente, di un teorema di Phillips sui semi-
gruppi di Hille, Zaidman ha dimostrato la quasi-periodicità delle soluzioni Y(x,t) a traiettoria relativamente compatta, supponendo f(x,t) dotata di derivata continua $f_t(x,t)$. Nel lavoro ci-
tato in [2], ho eliminato questa ipotesi, ma non quella di compattezza. Successivamente ho potuto rimuovere anche quest'ultima
ipotesi (molto restrittiva e di non chiara interpretazione fisica) pervenendo al seguente teorema [9] : *sia f(x,t) q.p., a valori in* L^2, *e sia Y(x,t) una soluzione dell'equazione variazionale
delle onde, limitata in J . Allora Y(x,t) è q.p.,come funzione a
valori in E .*

La dimostrazione di questo teorema è stata ottenuta in due
modi distinti, di cui ora indicheremo le linee fondamentali.

a) Nella prima dimostrazione (L.AMERIO, loc. cit. in [9])
ci si vale, in modo essenziale, di una proposizione sulle successioni monotone di funzioni q.p., proposizione che presenta analogia con quella, notissima, del Dini.

Sia, dapprima, $\{f_n(t)\}$ una successione di funzioni q.p., reali o complesse. Diciamo A la famiglia di tutte le successioni reali $a = \{a_k\}$ *regolari* rispetto a $\{f_n(t)\}$: tali cioè che, per ogni
n, sia

$$(3.1) \qquad \lim_{k \to \infty} f_n(t + a_k) = f_{n,a}(t) ,$$

uniformemente in J . La nuova successione $\{f_{n,a}(t)\}$ sarà forma-
ta, come la precedente, da funzioni q.p.

Inoltre, *una arbitraria successione reale* $\beta = \{\beta_k\}$ *contiene
una sottosuccessione* $a \in$ A (la tesi segue immediatamente dal cri-

L.Amerio

terio di Bochner, applicando il procedimento diagonale di Hilbert).

Supponiamo, ora, che la *successione* $\{f_n(t)\}$ *sia reale, mono-tona, e limitata indipendentemente da* t : risulti, ad esempio, in J

$$(3.2) \qquad f_n(t) \geq f_{n+1}(t) \geq m > -\infty .$$

Si osservi che, presa comunque $a \in A$, la successione $\{f_{n,a}(t)\}$ gode delle medesime proprietà : se infatti vale la (3.2), risulta per la (3.1),

$$(3.3) \qquad f_{n,a}(f) \geq f_{n+1,a}(t) \geq m .$$

Dalla (3.3), posto

$$f_n(t) = f_{n,o}(t) ,$$

si deduce che, per ogni $a \in A$, esiste il limite

$$(3.4) \qquad \lim_{n \to \infty} f_{n,a}(t) = F_a(t)$$

Ciò premesso, si dimostra che *la successione* $\{f_n(t)\}$ *converge uniformemente in J se le funzioni* $F_a(t)$ *risultano q.p., per ogni* $a \in A$. La dimostrazione del teorema dianzi enunciato si fonda allora sulle seguenti considerazioni.

I) Se è, per $t \in J$,

$$(3.5) \qquad \|Y(x,t)\|_E \leq M ,$$

si deduce dalle (2.6), (2.8) che gli integrali

$$\int_o^t f_n(\eta) \sin \lambda_n \eta d\eta , \qquad \int_o^t f_n(\eta) \cos \lambda_n \eta d\eta$$

sono limitati. Poichè $f_n(t)$ è q.p. (per la (2.10)) gli integrali medesimi risultano q.p. : tali sono allora le funzioni $\gamma_n(t)$, $\gamma'_n(t)$.

Per provare il teorema basta allora dimostrare che la serie (2.6) converge uniformemente in J , cioè che *la serie di funzioni*

L.Amerio

q.p. non negative

$$(3.6) \qquad \sum_{1}^{\infty} \left(\gamma_n^2(t) + \frac{\gamma_n'^2(t)}{\lambda_n^2} \right) = \| Y(x,t) \|_E^2$$

converge uniformemente in J .

II) Si dimostra che l'energia totale $\frac{1}{2} \| Y(x,t) \|_E^2$ è q.p.

Per questo basta osservare che, per la (2.12), risulta

$$\frac{d}{dt} \| Y(x,t) \|_E^2 = 2 \, (f(x,t), \, y_t(x,t))_{L^2} =$$

$$= 2 \sum_{1}^{\infty} f_n(t) \, \frac{\gamma_n'(t)}{\lambda_n}$$

e si riconosce che la serie a secondo membro converge uniformemente in J . $\| Y(x,t) \|_E^2$, limitata e dotata di derivata q.p., è allora q.p.

III) Sia $a = \{a_k\}$ una arbitraria successione reale. Si può, senz'altro, supporre a regolare rispetto alla successione

$$\left\{ \gamma_n^2(t) + \frac{\gamma_n'^2(t)}{\lambda_n^2} \right\} \quad \text{e tale che sia, uniformemente in J ,}$$

$$\lim_{k \to \infty} f(x, t + a_k) = g(x,t) \ ,$$

$$\lim_{k \to \infty} \gamma_n(t + a_k) = \bar{\gamma}_n(t), \quad \lim_{k \to \infty} \gamma_n'(t + a_k) = \bar{\gamma}_n'(t)$$

risultando

$$\bar{\gamma}_n(t) = \bar{a}_n \cos \lambda_n t + \bar{b}_n \sin \lambda_n t + \int_0^t g_n(\eta) \sin \lambda_n (t - \eta) d\eta \ ,$$

$$\sum_{1}^{\infty} \left(\bar{a}_n^2 + \bar{b}_n^2 \right) \leq M^2 \ ,$$

$$g_n(t) = \lim_{k \to \infty} f_n(t + a_k) = (g(x,t), u_n(x))_{L^2} \ .$$

La funzione

L.Amerio

$$Z(x,t) = \left\{ \sum_{1}^{\infty} \bar{\gamma}_n(t) \; \frac{u_n(x)}{\lambda_n} \; , \; \sum_{1}^{\infty} \frac{\bar{\gamma}'_n(t)}{\lambda_n} \; u_n(x) \right\}$$

è allora soluzione dell'equazione variazionale delle onde, con termine noto $g(x,t)$, e si ha

$$(3.7) \qquad \| Z(x,t) \|_E \leq M .$$

Ne segue, come in II), che la funzione

$$\| Z(x,t) \|_E^2 = \sum_{1}^{\infty} \left(\bar{\gamma}_n^2(t) + \frac{\bar{\gamma}'^2_n(t)}{\lambda_n^2} \right)$$

è q.p.. Di qui, per la proposizione, già enunciata, sulle successioni monotone di funzioni q.p., si deduce la convergenza uniforme, in J, della serie (3.6).

b) La seconda dimostrazione $^{(10)}$ utilizza, innanzi tutto, la nozione di *quasi-periodicità debole*. Dato uno spazio B, di Banach, qualsiasi, diciamo B^* lo spazio duale, formato dai funzionali lineari continui, definiti in B.

La funzione $y = f(t)$, a valori in B, si dirà allora *debolmente q.p. (d.q.p.) se per ogni $b^* \in B^*$ la funzione numerica $< f(t), b^* >$*

è q.p. Se B è hilbertiano, questo equivale ad ammettere che, *per ogni $b \in B$, il prodotto scalare $(f(t), b)_B$ risulti q.p.*

In ogni caso, si dimostra che *condizione necessaria e sufficiente perchè $f(t)$, d.q.p., sia q.p. è che la corrispondente traiettoria, in B, risulti relativamente compatta.*

Ciò premesso, la dimostrazione si svolge nel modo seguente.

I') Si riconosce, innanzi tutto, come in I), che se vale la (3.5), la funzione $Y(x,t)$ è d.q.p.

II') Si dimostra che, se vale la (3.5), presa ad arbitrio una successione reale $a = \{a_k\}$, si può estrarre da questa una sottosuccessione $a' = \{a'_k\}$ tale che la successione $\{Y(x,t + a'_k)\}$ converga de-

L.Amerio

bolmente e uniformemente in J a una funzione $Z(x,t)$, d.q.p. e soluzione dell'equazione variazionale delle onde, con termine noto

$$g(x,t) = \lim_{k \to \infty} f(x, t + a'_k) \ .$$

E' allora

$$\sup_{t \in J} \|Z(x,t)\|_E \leq \sup_{t \in J} \|Y(x,t)\|_E$$

e inoltre (siccome la successione $\{Z(x,t - a'_k)\}$ converge debolmente a $Y(x,t)$)

$$\sup_{t \in J} \|Y(x,t)\|_E \leq \sup_{t \in J} \|Z(x,t)\|_E \ .$$

Vale pertanto l'equaglianza

$$(3.7) \qquad \sup_{t \in J} \|Z(x,t)\|_E = \sup_{t \in J} \|Y(x,t)\|_E \ .$$

III') Si dimostra che $Y(x,t)$ ha traiettoria relativamente compatta,
provando che, se così non fosse, la $Z(x,t)$ definita in II') soddi-
sferebbe a una limitazione del tipo

$$\sup_{t \in J} \|Z(x,t)\| < \sup_{t \in J} \|Y(x,t)\|$$

ciò che è assurdo, per la (3.7).

Nella dimostrazione sono utilizzate, in modo essenziale : la
struttura hilbertiana dello spazio E (attraverso il teorema del pa-
rallelogramma), la costanza della norma delle soluzioni dell'equazione omogenea (principio di conservazione dell'energia), il teorema di dipendenza continua dal termine noto.

4.- La dimostrazione data in b) si può estendere all'equazione a-
stratta $^{(11)}$

$$(4.1) \qquad \frac{d\,u}{dt} = A(t)\,u(t) + f(t) \ .$$

L. Amerio

Nella (4.1) A(t) è una famiglia di operatori illimitati quasi-periodici (in un certo senso) ed f(t) una funzione q.p., a valori in uno spazio Y ; la soluzione u(t) si suppone a valori in uno spazio $X \subseteq Y$ uniformemente convesso e tale che l'immersione di X in Y sia continua.

In questo caso, assai generale, non si può più ammettere la quasi-periodicità delle soluzioni dell'equazione omogenea, nè la costanza, in J , della loro norma. E' possibile però pervenire ad una estensione del *secondo teorema* di Favard, facendo su tali soluzioni e su quelle delle equazioni associate

$$w'(t) = B(t) \, u(t)$$

(ove B(t) appartiene alla chiusura della famiglia $\{A(t + h)\}$) delle ipotesi di comportamento asintotico analoghe a quelle poste dal Favard : in sostanza la traiettoria di w(t) *non* deve avere lo zero come punto di accumulazione. Si dimostra allora che *se la* (4.1) *ha una soluzione limitata ne ha una quasi-periodica* (si tratta di soluzioni *deboli* e all'ipotesi di limitatezza si aggiunge quella, assai poco restrittiva, di uniforme continuità debole in J). Precisamente *risulta q.p. la soluzione (che esiste ed è unica) caratterizzata dalla condizione che sia minima la quantità*

$$\underset{t \in J}{\mathrm{Sup}} \ \|u(t)\|$$

(teorema di *minimax*). Questo teorema, già dimostrato per l'equazione delle onde (L. AMERIO, loc.cit. in [2]), ha, in tal caso, un significato fisico notevole, poichè afferma *l'esistenza di una e una sola soluzione per cui l'estremo superiore, in J , dell'energia abbia il più piccolo valore possibile.*

L.Amerio

Osserviamo che, in un assai recente lavoro [12], lo Zaidman ha
applicazione dei precedenti risultati, al problema di Cauchy per
quazione delle onde in *tutto lo spazio,* con termine noto q.p. :
tratta proprio di un problema in cui l'equazione omogenea ha del-
soluzioni limitate che non sono quasi-periodiche. Lo stesso Auto-
ha, infine, ampiamente studiato il problema delle soluzioni, che
no *distribuzioni quasi-periodiche,* delle equazioni iperboliche,
termine noto *distribuzione quasi-periodica.*

)
S.BOCHNER, *Abstrakte fastperiodische Funktionen,* Acta Math.,61
(1933).

)
O.A.LADYZENSKAIA, *Il problema misto per l'equazione iperbolica*
(in russo), Mosca Leningrado, 1953; B.Sz.
NAGY, *Vibrations d'une corde non homogène,*
Bull. Soc. Math. de France, 1947; L.AMERIO,
Problema misto e quasi periodicità per l'equa
zione delle onde non omogenea, Ann.di Mat.,
29, (1960).

)
C.F.MUCKENHOUPT, *Almost periodic functions and vibrating sy-*
stems, Journ. of Math. and Phys., MIT, 8 (1929

)
S.BOCHNER, *Fastperiodische Lösungen der Wellen-Gleichung,* Acta
Math., 62 (1934).

)
S.BOCHNER, J. von NEUMANN, *On compact solutions of operational*
differential equations, Ann. of Math
36 (1935).

)
S.SOBOLEV, *Sur la presque-périodicité des solutions de l'équa-*
tion des ondes, I, II, III, Comp. rend. Ac. Sc.
URSS, 1945.

L.Ameri⟨

(7)

    O.A.LADYZENSKAIA, loc. cit. in (2).

(8)

    S.ZAIDMAN, *Sur la presque-périodicité des solutions de l'équa-
tion des ondes non homogène*, Journ. of Math. and
Mech., 8 (1959).

(9)

    L.AMERIO, *Quasi-periodicità degli integrali ad energia limita-
ta dell'equazione delle onde, con termine noto quasi-
periodico*, I, II, III, Rend. Acc. Naz. dei Lincei, 28
(1960); cfr. S.BOCHNER, *Almost periodic solutions of
the in homogeneous Wave equation*, Proc. of the Nat.
Ac. of Sc. (1960).

(10)

    L.AMERIO, *Sull'equazione delle onde con termine noto quasi-pe-
riodico*, Rend. di Mat., 9 (1960).

(11)

    L.AMERIO, *Sulle equazioni differenziali quasi-periodiche a-
stratte*, Ricerche di Mat., 10 (1961).

(12)

    S.ZAIDMAN, *Solutions presque-périodiques dans le problème de
Cauchy, pour l'équation non homogène des ondes*,
Acc. Naz. dei Lincei (in corso di stampa).

(13)

    S.ZAIDMAN, *Solutions presque-périodiques des équations hyper-
boliques* (ancora da pubblicare).

CENTRO INTERNAZIONALE MATEMATICO ESTIVO

(C.I.M.E.)

L A W R E N C E   M A R K U S

SISTEMI DINAMICI CON STABILITA' STRUTTURALE

ROMA – Istituto Matematico dell'Università – 1960

$$\text{SISTEMI DINAMICI CON STABILITA' STRUTTURALE}$$

di

LAWRENCE MARKUS

ESEMPI E DEFINIZIONI.

Consideriamo come esempio l'equazione differenziale

$$\ddot{x} + b\,\dot{x} + k^2 x = 0 \ , \qquad b \geq 0 \ , \ k^2 > 0 \ \text{ costanti reali}$$

ovvero il sistema del 1° ordine

$$\dot{x} = y \ , \quad \dot{y} = -k^2 x - b\,y \ .$$

Si tratta di un sistema differenziale nel piano reale R^2 od anche sulla sfera $S^2 = R^2 + \infty$ (mediante l'ordinaria proiezione stereografica).

Se $b = 0$ si ha l'oscillatore armonico. Ogni soluzione, qualunque siano le condizioni iniziali, è periodica. Si ha così una famiglia di curve chiuse intorno all'origine del piano.

Se $b > 0$ si ha l'oscillatore con attrito.

Nel caso $b = 0$ non vi è stabilità strutturale poichè un cambiamento della b , per quanto piccolo, cambia radicalmente la configurazione della famiglia delle curve integrali. Invece il caso $b > 0$ presenta la stabilità strutturale poichè un cambiamento sufficientemente piccolo di b in $b(x,y) > 0$ e analogamente per il coefficiente k^2 dà ancora un oscillatore con attrito.

Consideriamo, in generale, un sistema differenziale reale

$$S : \quad \dot{x}^i = f^i(x^1,\ldots,x^n) \ , \qquad i = 1,2,\ldots,n$$

con le $f^i(x^1,\ldots,x^n) \in C^1$ in $\ominus$, insieme aperto e limitato dello R^n euclideo (reale).

151

L.Markus

Per i teoremi di esistenza e di unicità per ogni punto di Θ
passa una curva integrale. Consideriamo la famiglia di tali curve.
Allora S è un campo di vettori in Θ ed una curva integrale è una
curva tangente a tale campo.

Sia $\mathcal{B}$ lo spazio di Banach che si ottiene prendendo tutti i sistemi differenziali come S . Precisamente $\mathcal{B}$ consista di tutti i
campi vettoriali di classe C^1 in Θ e sulla frontiera $\partial\Theta$ (che
supporremo regolare, ad esempio una varietà differenziabile di dimensione n-1). La distanza tra due campi vettoriali S ed S' :
$$\dot{x}^i = g^i(x^1,\dots,x^n) \text{ sia}$$

$$\|S - S'\| = \max_{x \in \overline{\Theta},i} |f^i(x) - g^i(x)| + \max_{x \in \overline{\Theta},i,j} \left|\frac{\partial f^i}{\partial x^j} - \frac{\partial g^i}{\partial x^j}\right|$$

Def. Un sistema differenziale $S \in \mathcal{B}$ ha la stabilità strutturale se

    1. S non è tangente alla frontiera $\partial\Theta$ (S non è nullo su $\partial\Theta$);

    2. Per ogni $\epsilon > 0$ esiste un $\delta > 0$ tale che quando $\|S - S'\| < \delta$
per $S' \in \mathcal{B}$ allora S ed S' sono ϵ-omeomorfi, vale a dire esiste
un'applicazione topologica di $\overline{\Theta}$ su $\overline{\Theta}$ la quale porta la famiglia
delle curve integrali (non parametrizzate) di S sulla famiglia di
quelle di S', in modo tale che la distanza di ogni punto di Θ
dalla sua immagine è $\leq \epsilon$.

Allora se si cambiano di poco, le f^i , la configurazione glo-
bale delle curve integrali non muta qualitativamente.

Il concetto di stabilità strutturale è molto interessante dal
punto di vista filosofico poichè un sistema dinamico della fisica
non deve cambiare qualitativamente allorchè si produce un piccolo
cambiamento nei coefficienti. Ma l'idea è difficile a trattarsi dal
punto di vista matematico.

Se l'insieme Θ è aperto e limitato in una varietà M^n (invece
che in R^n) la teoria procede in modo analogo. In quel che segue sa-

152

L.Markus

rà considerato il caso più semplice $\Theta = M^n$, varietà differenziabile compatta. In tal caso la frontiera $\partial \Theta = \phi$ (insieme vuoto) e
la condizione 1. nella def. è sempre verificata.

SCHEMA DELLA TEORIA.

La teoria dei sistemi strutturalmente stabili si può descrivere
nei seguenti punti : 1. Esistenza; 2. Densità in $\mathcal{B}$; 3. Proprietà;
4. Applicazioni.

Ci soffermeremo in particolare sui due ultimi e accenniamo ra-
pidamente ai primi due.

L'esistenza di sistemi strutturalmente stabili è stata provata
nel caso di un disco da Andronov e Pontriaghin (v. Bibliografia) e
la prova si può estendere ad ogni M^2 : per M^3 si veda la comunicazio-
ne di S.Smale citata nella Bibliografia. Cosa può dirsi per M^4?

Quanto alla densità in $\mathcal{B}$, vale a dire al fatto che i sistemi
strutturalmente stabili sono densi nello spazio $\mathcal{B}$ di tutti i sistemi, essa risulta finora provata solo in M^2 (vedi De Baggis e Pei-
xoto). Cosa può dirsi in M^3?

PROPRIETA'.

Elenchiamo adesso alcune proprietà relative ai sistemi strutturalmente stabili in una varietà differenziabile compatta M^n.

1. Esiste un numero finito di punti singolari (punti d'equili-
brio) ed ognuno di essi è un punto elementare. (Ciò significa che
se P è un punto singolare, cioè un punto in cui il vettore del campo è nullo, esiste un'applicazione topologica di un intorno di P
su di un intorno dell'origine di R^n che porta le soluzioni di S su
quelle di una equazione lineare a coefficienti costanti

$$\dot{x}^i = a^i_j x^j$$

L.Markus

tale che gli autovalori della matrice (a^i_j) hanno tutti parti reali diverse da zero e sono a due a due distinti).

Per provarlo, in R^n, approssimiamo S mediante un sistema

$$S' \; : \quad \dot{x}^i = P^i(x^1,\ldots,x^n) \; , \qquad i = 1, 2,\ldots,n$$

dove i $P^i(x)$ sono polinomi reali. Se i coefficienti dei P^i sono generici (nel senso della geometria algebrica) i punti singolari sono isolati. Nell'intorno di P scriviamo

$$S' \; : \quad \dot{x}^i = a^i_j \, x^j + Q^i(x)$$

Facciamo un'altra approssimazione, con $Q^i \equiv 0$ nell'intorno di P . Per l'ipotesi di stabilità strutturale S è qualitativamente lo stesso che $\dot{x}^i = a^i_j x^j$. La dimostrazione è così completa.

2. Le soluzioni periodiche sono isolate ed elementari. (Ciò vuol dire che per ogni soluzione periodica esiste un intorno tubolare N nel quale non vi sono soluzioni periodiche oltre quella considerata. Inoltre gli esponenti caratteristici della soluzione sono a due a due distinti e, salvo quella banale, hanno tutti modulo $\neq 1$).

Una questione importante per ora aperta è : esiste un numero finito soltanto di soluzioni periodiche?

3. Se s è una soluzione positivamente stabile secondo Poisson (avente cioè intersezione non vuota con l'insieme dei propri punti ω-limite) allora s è un punto singolare oppure una soluzione periodica o infine esiste un'infinità di soluzioni periodiche asintotiche ad s . (Congettura di Andronov).

4. Se S ammette una opportuna misura invariante su M^n allora esiste un'infinità di soluzioni periodiche.

Segue che un sistema di Hamilton con un numero finito di solu-

L.Markus

zioni periodiche non può essere strutturalmente stabile.

5. Ogni traiettoria che non sia nomade è una traiettoria centrale.

Per chiarire questo punto ricordiamo che secondo la teoria di Birkhoff una traiettoria s di S si dice nomade (wandering) se esiste un intorno tubolare di s che a partire da un certo istante non interseca mai più la propria immagine. L'insieme Θ_1 dei punti appartenenti a traiettorie nomadi è aperto in M^n, quindi il complementare $M_1 = M^n - \Theta_1$, costituito dai punti non nomadi è chiuso e quindi (essendo M^n compatta) compatto. Ripetendo la costruzione a partire dal sistema dinamico considerato in M_1, con la topologia relativa, si ottiene una successione decrescente di insiemi compatti

$$M^n \supset M_1 \supset M_2 \supset M_3 \supset \ldots$$

la cui intersezione M_r, non vuota, costituisce quello che Birkhoff chiama centro del sistema dinamico. In esso si trovano i punti singolari, le soluzioni periodiche e in generale le soluzioni stabili secondo Poisson. La costruzione di M_r è assai complicata in generale. L'affermazione fatta all'inizio di questo n° è la seguente : Se il sistema S è strutturalmente stabile occorre un solo passo per ottenere il centro poichè $M_1 = M_r$.

APPLICAZIONI.

1. Consideriamo il flusso lungo le geodetiche d'una superficie chiusa a curvatura negativa costante. Questione : Un tale sistema dinamico ha stabilità strutturale?

2. Problema di H.Seifert.

Consideriamo un campo di vettori sulla sfera S^3, senza punti singolari. Esiste sempre una soluzione periodica? Nel caso della stabilità strutturale la risposta è affermativa. Infatti se si segue

L.Markus

una traiettoria il suo insieme ω-limite è una traiettoria periodica oppure è il limite d'una successione di traiettorie periodiche.

3. Generalizzazione del teorema di Poincaré-Bendixon.

Dato in un toro solido di R^3 un campo di vettori senza punti
singolari tali che quelli associati alla superficie penetrino tutti nell'interno esiste una traiettoria periodica? La risposta è
affermativa nel caso che vi sia stabilità strutturale.

L.Markus

BIBLIOGRAFIA

1. A.Andronov - L.Pontrjagin, *Systèmes grossiers*, Dokl. Akad. Nauk SSSR, 14, 247-251 (1937).

2. G.D.Birkhoff, *Dynamical Systems*, New York (1927).

3. H.De Baggis, *Dynamical systems with stable structures*, Contrib. to the Theory of nonl. oscill., 2, 37-59, Princeton (1952).

4. L.Markus, *Global structure of ordinary differential equations in the plane*, Trans. Am. Math. Soc. 76, 127-148 (1957).

5. L.Markus, *Structurally stable differential systems*, Notices of the Am. Math. Soc., (1959).

6. L.Markus, *Periodic solutions and invariant sets of structurally stable differential equations*, Proc. of the Congress on Nonl. diff. eq., Mexico (1959).

7. L.Markus, *Invariant sets of structurally stable differential systems*, Proc. Nat. Acad. Sci. USA (1959).

8. L.Markus, *Structurally stable differential systems*, Ann. of Math. (1961).

9. M.M.Peixoto, *On structural stability*, Ann. of Math., 69, 199-222 (1959).

10. M.M.Peixoto, *Some examples on n-dimensional structural stability*, Proc. Nat. Acad. Sci. USA, 49, 633-636 (1959).

11. G.Sansone - R.Conti, *Equazioni differenziali non lineari*, Roma (1956).

12. H.Seifert, *Closed integral curves in 3-space and isotropic two-dimensional deformations*, Proc. Am. Math. Soc., 1, 287-302 (1950).

13. S.Smale, *Structurally stable differential systems on closed 3-manifolds*, Proc. of Congress on nonl. diff. eq., Mexico (1959).

CENTRO INTERNAZIONALE MATEMATICO ESTIVO

(C.I.M.E.)

G. PRODI

TEOREMI ERGODICI PER LE EQUAZIONI DELLA IDRODINAMICA

ROMA - Istituto Matematico dell'Università - 1960

TEOREMI ERGODICI PER LE EQUAZIONI DELLA IDRODINAMICA [1]

di G. PRODI

In questi ultimi anni è stato dimostrato ([4] e, successivamente, [8]) che il problema misto (nel senso di Hadamard) per le
equazioni di Navier-Stokes in due variabili spaziali è univocamente
risolubile "in grande".

Questo risultato permette di impostare problemi di comportamento asintotico per t → +∞ : stabilità, esistenza di soluzioni periodiche, teoremi ergodici. Si tratta dei problemi che hanno il maggior
interesse anche dal punto di vista fisico. In quello che segue svol-
geremo qualche considerazione su questi problemi; in molti punti
non potremo che segnalare e mettere in evidenza le difficoltà che
si incontrano.

Esaminati i (pochi) dati che l'analisi attualmente ci fornisce
intorno alle equazioni di Navier-Stokes richiameremo alcune nozioni
di carattere generale relative all'esistenza di misure invarianti.
Cercheremo poi di adattare al nostro caso la teoria di Kriloff e
Bogoliuboff [3] ed enunceremo alcuni risultati che si possono così
ottenere.

§ 1.- Indichiamo con Ω un insieme aperto limitato del piano, con
u un vettore reale di componenti u_j (j = 1,2), funzioni del punto
$x \in \Omega$ (x ≡ $(x_1 x_2)$).

---

(1)
    Le ricerche qui esposte sono state finanziate dall'Air Research
and Development Command, United States Air Force, con contratto
AF 61 (052) - 414.

G. Prodi

Indicheremo con $L^p(\Omega)$ lo spazio dei vettori con componenti a p-esima potenza sommabile, con la norma $|u|_{L^p(\Omega)} = \{\int_\Omega |u(x)|^p \, dx\}^{1/p}$, dove $|u(x)|$ indica il modulo di $u(x)$. Se $p = 2$, questo spazio ha struttura hilbertiana (reale), individuabile mediante il prodotto scalare $(f, g) = \int_\Omega \sum_i f_i g_i \, dx$. Scriveremo $|f|$ in luogo di $|f|_{L^2(\Omega)}$.

Siano poi, f, g vettori definiti in Ω aventi derivate prime (in senso generalizzato) a quadrato sommabile in Ω. Scriveremo

$$((f,g)) = \int_\Omega \sum_{ij} \frac{\partial f_i}{\partial x_j} \frac{\partial g_i}{\partial x_j} \, dx \ , \ \|f\|^2 = ((f,f)).$$

Sia $\mathcal{N}(\Omega)$ la varietà dei vettori indefinitamente differenziabili, a divergenza nulla e nulli fuori di un compatto contenuto in Ω. Sia $N(\Omega)$ la chiusura di $\mathcal{N}(\Omega)$ secondo la norma $|\ |$, $N^1(\Omega)$ la chiusura di $\mathcal{N}(\Omega)$ secondo la norma $\|\ \|$. Gli elementi di $N^1(\Omega)$ sono, in un certo senso, vettori nulli sulla frontiera di Ω.

Indichiamo con C_Ω il coefficiente di immersione :

$$C_\Omega = \sup_{u \in N^1(\Omega)} |u| \ \|u\|^{-1}.$$

Ricordiamo la importante disuguaglianza di Ladyzenskaia (valida per un campo Ω qualsiasi) : $|u|^2_{L^4(\Omega)} \leq 2^{1/2} |u| \ \|u\|$ (vedi [4]).

Se u, v, $w \in N^1(\Omega)$ poniamo $b(u,v,w) = \int_\Omega \sum_{ij} u_i \frac{\partial v_j}{\partial x_i} w_j \, dx$.

Si ha facilmente $|b(u,v,w)| \leq |u|_{L^4(\Omega)} \|v\| \ |w|_{L^4(\Omega)}$.

Dato uno spazio di Banach B, indicheremo infine con $L^p(0,\tau;B)$ lo spazio delle funzioni definite nell'intervallo $(0,\tau)$, con valori in B, a p-esima potenza sommabile.

Le equazioni che considereremo con le relative condizioni al contorno, si esprimono, in termini classici, in questo modo:

G.Prodi

$$(1) \qquad \frac{\partial u_j}{\partial t} + \sum_i u_i \frac{\partial u_j}{\partial x_i} - \mu \, \Delta_2 \, u_j = - \frac{\partial p}{\partial x_j} + f_j \quad (j = 1,2)$$

$$(2) \qquad \text{div } u = 0$$

$$(3) \qquad u(x) = 0 \quad \text{per} \quad x \in \mathcal{F}r(\Omega) \; .$$

Qui la incognita p ha il significato di pressione, μ indica il coefficiente di viscosità. *Nel seguito noi supporremo sempre* $f = (f_1, f_2)$ *funzione della sola* x . Questo per metterci in una situazione di analogia, per quanto è possibile, con la teoria dei sistemi autonomi ordinari.

Supporremo $f \in L^2(\Omega)$; diremo che una funzione u(t), localmente limitata come funzione con valori in $N(\Omega)$, localmente a quadrato sommabile come funzione con valori in $N^1(\Omega)$, è soluzione (debole) delle equazioni di Navier-Stokes se vale la relazione

$$(4) \qquad \int_0^\infty \{-(u(t),v'(t)) + \mu((u(t),v(t))) + b(u(t),u(t);v(t))\} \, dt =$$
$$- \int_0^\infty (f(t),v(t)) \, dt$$

per ogni funzione v(t) che :

(a) sia continua come funzione di t con valori in $N^1(\Omega)$;

(b) si annulli fuori di un intervallo $\tau' \vdash \tau''$, con $0 < \tau' < \tau'' < +\infty$;

(c) abbia derivata rispetto a t $v'(t) \in L^2(0, +\infty; N(\Omega))$.

Si dimostra che :

1) (Vedi [9]) le soluzioni della (4), eventualmente corrette su un insieme di valori di t di misura nulla, sono funzioni continue di t nello spazio $N(\Omega)$.

2) (Vedi [9]) la funzione $|u(t)|^2$ è assolutamente continua e soddisfa all'equazione differenziale

$$(5) \qquad \frac{1}{2} \frac{d}{dt} |u(t)|^2 + \mu \|u(t)\|^2 = (f,u(t))$$

G. Prodi

3) (Vedi [7], [8]) Assegnato comunque un valore $u_0 \in N(\Omega)$, esiste
una ed una sola soluzione che lo assume per $t = 0$; essa è definita
in tutto l'intervallo $0 \longmapsto \infty$.

Posto $u(t) = T_t u_0$, T_t definisce una applicazione continua:
$(0 \longmapsto \infty) \times N(\Omega) \to N(\Omega)$. (Non mi è riuscito di vedere se la continuità in questione possa sussistere in modo uniforme).

Dalla (5) si ricava facilmente

$$(6) \qquad |u(t)| \leq |u_0| \, e^{-\frac{\mu}{c_\Omega^2} t} + c_\Omega^2 \, \mu^{-1} \, |f| \quad .$$

Si ha dunque, per ogni soluzione, la relazione

$$(7) \qquad \overline{\lim_{t \to +\infty}} \, |u(t)| \leq c_\Omega^2 \, \mu^{-1} \, |f| \, .$$

Ancora dalla (5) si ricava la diseguaglianza

$$\mu \int_0^T \|u(t)\|^2 \, dt \leq \frac{1}{2} \, |u_0|^2 + |f| \int_0^T |u(t)| \, dt \, .$$

Applicando la (6) si ha $^{(1)}$

$$(8) \qquad \int_0^T \|u(t)\|^2 \, dt \leq c_\Omega^2 \, \mu^{-2} \, |f| \, |u_0| + 2^{-1}\mu^{-1}|u_0|^2 + c_\Omega^2\mu^{-2}|f|^2 \tau \, .$$

Da questa si ottiene

$$(9) \qquad \lim_{\tau \to +\infty} \frac{1}{\tau} \int_0^T \|u(t)\|^2 \, dt \leq c_\Omega^2 \, \mu^{-2} \, |f|^2 \quad .$$

Aggiungiamo alcune osservazioni relative alle soluzioni costanti della (4). Un teorema di esistenza è stato dato recentemente dalla Ladyzenskaia [6]. (In ipotesi più restrittive e in altra impo-
stazione un teorema di esistenza era stato dato fin dal 1933 da
Leray).

----

(1)
    Diseguaglianze di questo tipo sono state introdotte da E. Hopf
in [1].

G.Prodi

L'unicità della soluzione si ha certamente per valori di $|f|$ piccoli (come constateremo indirettamente tra poco). Per valori arbitrari di $|f|$ non si conosce nulla in proposito. Si è portati a ritenere che, in questo caso, il teorema di unicità non sussista; tuttavia non mi risulta che sia stato trovato alcun esempio a riguardo. Per le soluzioni costanti si ottiene immediatamente dalla (9) : $\|u\| \leq \dfrac{C_\Omega}{\mu} |f|$. Fissata f , le soluzioni descrivono dunque un insieme S limitato in $N^1(\Omega)$ e perciò (per il teorema di Rellich) compatto in $N(\Omega)$.[1]

Consideriamo brevemente il problema della stabilità delle soluzioni costanti. Detta u una soluzione costante, $u^*(t) = u + w(t)$ un'altra qualsiasi soluzione, dalla (4), con considerazioni analoghe a quelle che si fanno per ottenere la (5), si ricava :

$$(10) \qquad \frac{1}{2} \frac{d}{dt} |w(t)|^2 + \mu\|w(t)\|^2 + b(w(t),u,w(t)) = 0 .$$

Si ha

$$|b(w(t),u,w(t))| \leq \|u\| \, |w|^2_{L^4(\Omega)} \leq 2^{1/2} \|u\| \, \|w\| \, |w| \leq 2^{1/2} C_\Omega \|u\| \, \|w\|$$

Supponiamo ora che sia $2^{1/2} C_\Omega \mu^{-1} \|u\| < 1$ (il numero $2^{1/2} C_\Omega \mu^{-1} \|u\|$ differisce per una costante da quello che è detto "numero di Reynolds generalizzato" dalla Ladyzenskaia [5]).

Ponendo $2^{1/2} C_\Omega \|u\| = \mu - \epsilon$, otteniamo $\mu\|w(t)\|^2 + b(w,u,w) \geq \epsilon \|w(t)\|^2$.

Si ha dunque dalla (10)

---

[1]
Si può dimostrare che l'insieme S delle soluzioni costanti ha questa proprietà : per ogni suo punto u_0 esiste una varietà lineare ad un numero finito di dimensioni che è, in un certo senso, "tangente" ad S (in $N^1(\Omega)$); in un intorno di u_0 , S si proietta ortogonalmente (sempre in $N^1(\Omega)$) su questa varietà senza sovrapposizioni. Non è detto però che questa proiezione sia "su".

G.Prodi

$$\frac{1}{2} \, \frac{d}{dt} \, |w(t)|^2 + \epsilon \, c_\Omega^{-2} \, |w(t)|^2 \le 0 \, .$$

Da questa è facile dedurre che, per $t \to +\infty$, $w(t) \to 0$, e che $\int_o^{+\infty} \|w(t)\|^2 \, dt < +\infty$. Dunque, se è $2^{1/2} c_\Omega \, \mu^{-1} \|u\| < 1$ (il che accade se è $|f| < 2^{-1/2} \, \mu^2 c_\Omega^{-2}$) la soluzione u è stabile; segue anche che essa è unica.

Sul problema della stabilità delle soluzioni costanti non si conosce nulla più di queste semplici osservazioni che riguardano soltanto il caso di u "piccolo" (e che si trovano in [5]). Il problema della stabilità è legato strettamente a quello dell'unicità; vedremo successivamente come questo problema è legato al problema che ci interessa principalmente.

§2.- Richiameremo qui alcuni risultati relativi alla teoria della misura sugli spazi metrici e alla teoria ergodica classica, allo scopo principale di adattarli al nostro problema.

Sia E uno spazio metrico separabile completo, sia $\mathcal{B}(E)$ la famiglia dei boreliani di E . Si dice misura di probabilità una funzione non negativa definita per un σ-campo $\mathcal{F}$ di insiemi di E , σ-additiva, tale che $E \in \mathcal{F}$ e $m(E) = 1$. *Noi intenderemo sempre, inoltre che m sia il completamento alla Lebesgue di una misura definita in $\mathcal{B}(E)$.*

Dalla misura m si ottiene un integrale, con la consueta definizione. Porremo $M(f) = \int_E f(u) \, dm(u)$.

Indichiamo con $\mathcal{C}(E)$ lo spazio delle funzioni continue e limitate definite in E , con la norma $|f|_{\mathcal{C}(E)} = \sup_{u \in E} |f(u)|$. L'integrale definisce dunque in $\mathcal{C}(E)$ un funzionale $M(f)$ lineare, continuo non negativo e che, inoltre, gode di questa proprietà (di Daniell) : per ogni successione $\{f_h\}$ di funzioni continue non

G.Prodi

negative convergenti a zero in modo monotono si ha : $\lim_{n \to \infty} M(f_n) = 0$.

Viceversa supponiamo di assegnare in $\mathcal{C}(E)$ un funzionale che goda

di queste proprietà (diremo: un integrale astratto); esso si può

prolungare con un procedimento ben noto [1] ; si potranno allora dire

misurabili gli insiemi le cui funzioni caratteristiche risultano

sommabili. Se aggiungiamo la condizione di normalizzazione : $M(1) = 1$

(con 1 indichiamo qui la funzione che è uguale ad 1 su E), abbiamo

una completa equivalenza tra misura (di probabilità) ed integrale.

Partendo da una misura m e definendo un integrale M, attraverso que-

sto riotteniamo la medesima misura di partenza. Così, partendo da

un integrale M e passando agli insiemi misurabili, possiamo riotte-

nere l'integrale.

La considerazione di un integrale al posto di una misura, è

spesso più comoda. Questo vale ad esempio, per istituire una topo-

logia nell'insieme della misura. Noi doteremo l'insieme delle misure

della topologia che viene subordinata in esso dalla topologia debole

dello spazio $\mathcal{C}^*$. (Identificando le misure con i relativi integrali,

che sono elementi di $\mathcal{C}^*$). E' importante avere criteri di compattez-

za per le misure. La sfera unitaria dello spazio $\mathcal{C}^*(E)$ è, in ogni

caso, compatta (sempre nella topologia debole). Ora, se lo spazio

E è compatto, tutti i funzionali non negativi sono misure (infatti,

in virtù del teorema di Dini, sussiste la proprietà di Daniell).

Poichè la sfera unitaria dello spazio $\mathcal{C}^*(E)$, nella topologia debo-

le, possiede una base numerabile di intorni, si deduce che *da ogni*

successione $\{m_n\}$ *di misure di probabilità si può estrarre una suc-*

cessione parziale convergente verso una misura di probabilità.

---

(1)
    Vedere ad esempio, Loomis, Abstract harmonic analysis, (1953)
Cap.III.

167

G.Prodi

Su questo risultato si fonda la dimostrazione dell'esistenza di una misura invariante rispetto ad un gruppo $\{T_t\}$ di trasformazioni $(-\infty < t < +\infty)$ data da Kryloff e Bogoljuboff [3].

Noi abbiamo bisogno di estendere, in qualche modo, il risultato enunciato al caso in cui E non sia compatto. Da un risultato di Ulam (richiamato in [11]) si ha che, assegnata in uno spazio E metrico, completo, separabile, una misura m , per ogni $\epsilon > 0$ si può trovare un compatto D tale che $m(E-D) \leq \epsilon$. Noi considereremo in E una famiglia di misure di probabilità che soddisfino a questa condizione *uniformemente*. Precisamente : sia assegnata una successione $\{D_n\}$ $(n = 1,2,3,\ldots)$ di insiemi compatti, con $D_n \subset D_{n+1}$. Sia $\mathcal{M}$ la famiglia di tutte le misure di probabilità m che soddisfano alla condizione : $m(E-D_n) \leq \dfrac{1}{n}$.

Si ha allora :

La famiglia $\mathcal{M}$ è compatta, nella topologia introdotta.

Basterà dimostrare che la famiglia $\mathcal{M}$ è un sottoinsieme chiuso di $\mathcal{C}^*(E)$. Indicando sempre con M il funzionale corrispondente alla misura m , la condizione che individua la famiglia $\mathcal{M}$ si traduce nella seguente : preso un intero $n > 0$, per ogni funzione ϕ , continua, nulla in D_n , con $0 \leq \phi \leq 1$, si ha $M(\phi) \leq \dfrac{1}{n}$.

Sia ora F un elemento di $\mathcal{C}^*(E)$ appartenente alla chiusura di $\mathcal{M}$. Evidentemente F sarà non negativo, tale che $F(1) = 1$. Inoltre soddisferà alla condizione : $F(\phi) \leq \dfrac{1}{n}$, per ogni funzione ϕ (con $0 \leq \phi \leq 1$) nulla in D_n . Basterà verificare che F gode della proprietà di Daniell. Sia $\{\phi_n\}$ una successione monotona non crescente di funzioni limitate. Possiamo supporre $\phi_n \leq 1$. Fissato un $\epsilon > 0$ e preso un intero m tale che $\dfrac{1}{m} < \dfrac{\epsilon}{2}$, consideriamo il dominio D_m . Ciascuna delle funzioni ϕ_n può essere decomposta in

G.Prodi

questo modo : $\phi_n = \phi_n' + \phi_n''$, dove $\phi_n' \geq 0$, $\phi_n'' \geq 0$, $\phi_h' = \phi_n$ sull'insieme D_m . Imporremo inoltre la condizione : $\|\phi_n'\|_{\mathcal{C}(E)} =$ $= \|\phi_n\|_{\mathcal{C}(D_m)}$. Poichè, per il teorema di Dini, ϕ_n converge uniformemente a zero su D_m esiste un intero n_o tale che, per ogni $n > n_o$ si ha $\|\phi_n'\|_{\mathcal{C}(E)} = \|\phi_n\|_{\mathcal{C}(D_m)} < \frac{\epsilon}{2}$. Per $n > n_o$ si avrà allora

$$F(\phi_n) = F(\phi_n') + F(\phi_n'') < \epsilon$$

Si ha inoltre

Nella topologia indotta in $\mathcal{M}$ da $\mathcal{C}^*(E)$ ogni punto possiede una base numerabile di intorni.

Infatti, sia $\{f_{nk}\}$ $(n,k = 1,2,\ldots)$ un sistema di funzioni continue soddisfacenti a queste condizioni : a) $\|f_{nk}\| \leq 1$; b) per ogni n fissato le restrizioni di f_{nk} a D_n $(k = 1,2,\ldots)$ costituiscono un insieme denso nella sfera unitaria di $\mathcal{C}(D_n)$; c) $\|f_{nk}\|_{\mathcal{C}(E)} = \|f_{nk}\|_{\mathcal{C}(D_n)}$.

Se $m^* \in \mathcal{M}$, gli insiemi formati dalle misure $m \in \mathcal{M}$, tali che $|(M - M^*)(f_{nk})| \leq \frac{1}{r}$ $(n,k,r = 1,2,\ldots)$ costituiscono un sistema fondamentale di intorni.

Possiamo concludere :

nelle ipotesi fatte, da ogni successione $\{m_n\}$ di misure appartenenti alla famiglia $\mathcal{M}$ si può estrarre una successione parziale convergente.

Supponiamo ora che sia assegnata una famiglia $\{T_t\}$ di trasformazioni dello spazio E in sè. Per potere comprendere il caso che ci interessa, supporremo che essa sia definita per $0 \leq t < +\infty$ e che costituisca un *semigruppo*. (Tutto fa pensare che le trasformazioni T_t definite nel § 1 non abbiano inversa continua). Supporremo (cfr. § 1, proprietà 3)) che T_t definisca una applicazione continua $(0 \longmapsto +\infty) \times E \to E$. Supponiamo assegnata in E una misura

G. Prodi

di probabilità m . Essa viene mutata dalla T_t (per ogni valore t $\geq$ 0 fissato) in una misura m_t, secondo la relazione

$$\int f(u)dm_t = \int f(T_t u)dm$$

valida per ogni funzione $f \in \mathcal{C}(E)$.

Se una funzione f è integrabile secondo m_t , la f' così definita : f'(u) = f(Tu) è integrabile secondo la m . Diremo che m è *invariante, se, per ogni valore di* t *si ha* $m_t = m$.

Se R è un insieme misurabile e indichiamo con f_R la sua funzione caratteristica, si ha nell'ipotesi che m sia invariante $\int f_R(u)dm = \int f_R(T_t u)dm$. La funzione $f_R(T_t u)$ è integrabile ed è la funzione caratteristica dell'insieme $T_t^{-1}(R)$.

Dunque : *se* m *è invariante l'immagine reciproca secondo la* T_t *di ogni insieme misurabile è misurabile ed ha la stessa misura di questo.*

Se introduciamo nello spazio $(0 \vdash +\infty) \times E$ la misura prodotto della ordinaria misura per una misura invariante m , abbiamo che le immagini *reciproche* di insiemi misurabili secondo la trasformazione : $T_t u : (0 \vdash +\infty) \times E \to E$, sono misurabili.

Il classico teorema di Birkhoff, inizialmente dato per il caso di un *gruppo* di trasformazioni, si estende (cfr. F. Riesz [10]) al caso da noi considerato, di un semigruppo di trasformazioni; *sia* m *una misura invariante e sia* f *una funzione integrabile, allora il limite* $\lim\limits_{\tau \to \infty} \dfrac{1}{\tau}\int_0^\tau f(T_t u)dt$ *esiste per quasi tutti i valori di* u . *La funzione limite è invariante ed è integrabile.*

§3.- Diamo ora alcuni risultati relativi all'esistenza di misure di probabilità definite in $N(\Omega)$, invarianti rispetto al semigruppo di trasformazioni di $N(\Omega)$ in sè, introdotto nel §1.

G. Prodi

Indicheremo con Σ la sfera : $|u| \leq C_\Omega^2 \, \mu^{-1} |f|$, con D_n (n intero > 0) l'insieme compatto : $\|u\|^2 \leq n$.

Gli integrali che scriveremo si intenderanno estesi a $N(\Omega)$, se non è altrimenti precisato.

I.- *Per una qualsiasi misura di probabilità invariante m si ha* $m(\Sigma) = 1$.

Infatti, detta f una qualunque funzione continua, limitata non negativa che assuma valore 1 su Σ , si ha per ogni t > 0 $\int f(u)dm = \int f(T_t u)dm$, essendo m invariante. Perciò per ogni τ > 0

$$\int f(u)dm = \frac{1}{\tau} \int_0^\tau \{\int f(T_t u)dm\}dt = \int\{ \frac{1}{\tau} \int_0^\tau f(T_t u)dt\}dm \quad .$$

Teniamo ora presente che, in virtù della (7) si ha $\lim_{\tau \to \infty} \frac{1}{\tau} \int_0^\tau f(T_t u)dt = 1$. Si deduce

$$\int f(u)dm = \int\{\lim_{\tau \to \infty} \frac{1}{\tau} \int_0^\tau f(T_t u)dt\}dm = \int 1 \, dm = 1$$

ora

$$m(\Sigma) = \inf \int f(u)dm = 1 \quad .$$

II.- *Per una qualsiasi misura di probabilità invariante m si ha* $m(D_n) \geq 1 - n^{-1} C_\Omega^2 \, \mu^{-2} |f|^2$.

Indichiamo con f_{D_n} la funzione caratteristica dell'insieme D_n . Procedendo dapprima come per il teorema precedente, poi applicando il teorema di Birkhoff, si ha

$$m(D_n) = \int f_{D_n}(u)dm = \int f_{D_n}(T_t u)dm = \lim_{\tau \to \infty} \frac{1}{\tau} \int_0^\tau \{\int f_{D_n}(T_t u)dm\}dt =$$

$$= \int\{\lim_{\tau \to +\infty} \frac{1}{\tau} \int_0^\tau f_{D_n}(T_t u)dt\}dm$$

Indichiamo ora con $\phi_n(u, \tau)$ la misura (ordinaria) del sottoinsieme dell'intervallo $0 \leq t \leq \tau$ per cui si ha $\|T_t u\|^2 > n$ (cioè

171

G.Prodi

$T_t u \notin D_n$).

Si ha $\tau - \phi_n(u, \tau) = \int_o^\tau f_{D_n}(T_t u)dt$; perciò esiste il limite $\lim_{\tau \to +\infty} \frac{1}{\tau} \phi_n(u, \tau)$ per quasi tutti i valori di u . D'altra parte, applicando la (9) si ha

$$\lim_{\tau \to +\infty} \frac{n}{\tau} \phi_n(u, \tau) \leq \overline{\lim}_{\tau \to +\infty} \frac{1}{\tau} \int_o^\tau \|T_t u\|^2 dt \leq C_\Omega^2 \mu^{-2} |f|^2 .$$

Perciò

$$m(D_n) = \int \{\lim_{\tau \to +\infty} \frac{1}{\tau} \int_o^\tau f_{D_n}(T_t u)dt\}dm \geq \int \{1 - n^{-1} C_\Omega^2 \mu^{-2} |f|^2\}dm =$$

$$= 1 - n^{-1} C_\Omega^2 \mu^{-2} |f|^2 .$$

Possiamo ora chiederci se esistono misure di probabilità invarianti. Una prima risposta è ovvia: ogni misura che ha il supporto contenuto nell'insieme S delle soluzioni costanti è senz'altro invariante. Chiameremo misure banali le misure di questo tipo. Interesserà sapere se esistono misure invarianti non banali. Si intuisce subito che, nel caso in cui tutte le soluzioni siano asintotiche alle soluzioni costanti, non potranno esistere che misure banali. Precisamente: fissiamo un punto $u_o \in N(\Omega)$. Diremo (seguendo K. e B. [3]) che *la traiettoria* $T_t u_o$ *è asintotica ad un insieme* S se, detta f_A la funzione caratteristica di un qualunque insieme aperto contenente S , si ha $\lim_{\tau \to +\infty} \frac{1}{\tau} \int_o^\tau f_A(T_t u_o)dt = 1$. Ora si ha:

Se tutte le traiettorie sono asintotiche all'insieme S *delle soluzioni costanti, ogni misura invariante è necessariamente banale.*

La dimostrazione si ottiene con lo stesso ragionamento fatto per il Teorema I. Viceversa :

III.- *Se esiste una traiettoria che non sia asintotica all'insieme* S *esiste almeno una misura di probabilità invariante non ba-*

G. Prodi

nale.

Supponiamo dunque che esistano un punto u_o ed un insieme aperto A tali che A $\subset$ S e che

$$\varlimsup_{\tau \to +\infty} \frac{1}{\tau} \int_0^\tau f_A (T_t u_o)dt = 1 - \gamma \quad (\text{con } \gamma > 0) .$$

Detto A' l'insieme complementare di A , si ha

(11) $$\varlimsup_{\tau \to +\infty} \frac{1}{\tau} \int_0^\tau f_{A'} (T_t u_o) dt = \gamma .$$

Indichiamo con m_τ la misura definita dall'integrale

$$M_\tau (f) = \frac{1}{\tau} \int_0^\tau f(T_t u_o)dt .$$

Dimostriamo che questa famiglia soddisfa alle condizioni del criterio di compattezza introdotto nel § 1, almeno per $\tau \geq 1$. Consideriamo infatti una qualunque funzione continua ψ , (con $0 \leq \psi \leq 1$) nulla in D_n e teniamo presente la funzione $\phi_n (u_o, \tau)$, definita come nella dimostrazione del teorema precedente. Si ha, indicando con D'_n il complementare di D_n ,

$$M_\tau (\psi) = \frac{1}{\tau} \int_0^\tau \psi(T_t u_o)dt \leq \frac{1}{\tau} \int_0^\tau f_{D'_n} (T_t u_o) dt \leq$$

$$\leq \frac{1}{n} \frac{1}{\tau} \int_0^\tau \| T_t u_o \|^2 dt .$$

Applicando la (8) si ottiene dunque $m(N(\Omega) - D_n) =$

$$= \sup_\phi M_\tau (\phi) \leq \frac{1}{n} C_\Omega^2 \mu^{-2} |f|^2 + \frac{1}{n\tau} (C_\Omega^2 \mu^{-2}|f||u_o| + 2^{-1}\mu^{-1}|u_o|^2) .$$

Questo dimostra che le condizioni del criterio di compattezza sono soddisfatte.

Tenendo presente la (11) consideriamo una successione $\{\tau_k\}$ divergente a $+\infty$ e tale che

G.Prodi

$$(12) \quad \lim_{\tau_k \to +\infty} \frac{1}{\tau_k} \int_0^{\tau_k} f_{A'}(T_t u_0)\, dt = \gamma \quad .$$

Estraiamo dalla successione di misure $\{m_{\tau_k}\}$ una successione $\{m_{\tau_{k'}}\}$ convergente. Indichiamo con m la misura limite, con M il relativo integrale. Dimostriamo anzitutto che m non è banale. Sia infatti g una funzione continua, limitata non negativa, assumente valore 1 nell'insieme A'. Si avrà tenendo presente la (12),

$$M(g) = \lim_{\tau_{k'} \to +\infty} \frac{1}{\tau_{k'}} \int_0^{\tau_{k'}} g(T_t u_0)\, dt \geq \lim_{\tau_{k'} \to +\infty} \frac{1}{\tau_{k'}} \int_0^{\tau_{k'}} f_{A'}(T_t u_0)\, dt = \gamma$$

Perciò $m(A') = \inf_{g} M(g) \geq \gamma$.

La misura m è poi invariante. Ripetendo infatti il ragionamento di K. e B. [3], si ha (indicando con $T_\eta f$ la funzione $T_\eta f(u) = f(T_\eta u)$)

$$M(T_\eta f) = \lim_{\tau_{k'} \to +\infty} \frac{1}{\tau_{k'}} \int_0^{\tau_{k'}} f(T_{\eta+t} u_0)\, dt =$$

$$= \lim_{\tau_{k'} \to +\infty} \frac{1}{\tau_{k'}} \int_\eta^{\tau_{k'}+\eta} f(T_t u_0)\, dt = M(f) \quad .$$

Resta dunque il problema di *dare criteri che stabiliscono quando le soluzioni sono asintotiche all'insieme S delle soluzioni costanti e quando si verifica il caso contrario.*

Da una misura di probabilità invariante non banale si può ovviamente ottenere una misura di probabilità invariante per cui $m(S) = 0$.

Si pone allora il problema (che si potrebbe ancora dire della *ergodicità* della trasformazione) : *una tale misura è unica?*

Sarebbe anche interessante indagare su una congettura formulata da E.Hopf [2], congettura che possiamo tradurre in questi termini : il supporto di una misura invariante non banale è una

174

G.Prodi

varietà di dimensione finita. La dimensione cresce con il decrescere del coefficiente di viscosità.

Anche prescindendo dalla risoluzione di questi problemi, seguendo K. e B. [3] , e apportando lievi modifiche ai ragionamenti da loro svolti, possiamo ottenere interessanti risultati. Ci limitiamo ad accennarne qualcuno a titolo di esempio.

Secondo K. e B. si definisce insieme di probabilità nulla un insieme di misura nulla rispetto ad una qualsiasi misura invariante; si dice insieme di probabilità massima un insieme il cui complementare ha probabilità nulla.

Allora, come immediata conseguenza del teorema di Birkhoff, si ha : *se f è una funzione continua, l'insieme dei punti u per cui il limite* $\lim\limits_{\tau \to +\infty} \dfrac{1}{\tau} \int_0^\tau f(T_t u)dt$ *non esiste è di probabilità nulla.*

Ad esempio, se prescindiamo da un insieme di valori iniziali u_0 di probabilità nulla, le soluzioni $u(t)$ sono tali che esiste il limite $\lim\limits_{\tau \to +\infty} \dfrac{1}{\tau} \int_0^\tau |u(t)|^p dt$ per ogni $p > 0$.

Se diciamo *quasi regolari* i punti u per cui il limite $\lim\limits_{\tau \to +\infty} \dfrac{1}{\tau} \int_0^\tau f(T_t u)dt$ esiste qualunque sia la funzione continua f , si ha : *l'insieme dei punti quasi regolari è invariante ed ha probabilità massima.*

G.Prodi

B I B L I O G R A F I A

[1] E. HOPF - "Ein allgemeiner Endlichkeitssatz der Hydroaynamik".
     Math.Annalen 117, 764-775 (1941).

[2] E. HOPF - "A mathematical example displaying features of tur-
     bulence". Comm. on pure and appl. math. 1, 303-322
     (1948).

[3] N.KRYLOFF e N.BOGOLJUBOFF - "La théorie générale de la mesu-
     re dans son application a l'étude des systèmes dy-
     namiques de la mécanique non linéaire". Annals of
     Math. 38, 65-113 (1937).

[4] O.A. LADYZENSKAIA - "Soluzioni in grande del problema al con-
     torno non stazionario, per il sistema di Navier-
     Stokes in due variabili spaziali". Dokl. Akad. Nauk.
     123 (3) 1128-1131 (1958). (in Russo)

[5] O.A. LADYZENSKAIA - "Soluzioni in grande del problema al con-
     torno non stazionario, per il sistema di ....
     Comm. on Pure and Appl. Math. 12, 427-433 (1959).

[6] O.A. LADYZENSKAIA - "Lo studio delle equazioni di Navier-Sto-
     kes nel caso del movimento stazionario di un fluido
     incompressibile". (in Russo)
     Uspehi Math. Nauk 14, (1959).

[7] J.L. LIONS - "Quelques résultats d'existence dans des équa-
     tions aux dérivées partielles non linéaires".
     Bull.Soc.Math. France, 87, 245-273 (1959).

[8] J.L. LIONS e G. PRODI - "Un théorème d'existence et unicité
     dans les équations de Navier-Stokes en dimension 2".
     Comptes rendus des séances de l'Acad. des Sciences,
     t. 248, p.3519-3521.

[9] G. PRODI - "Qualche risultato riguardo alle equazioni di Na-
     vier-Stokes nel caso bidimensionale". Rend.Sem.Mat.
     Padova, 30, 1-15 (1960).

[10] F. RIESZ - "Sur la théorie ergodique". Comm.Math. Helvetici,
     17, 221-239 (1944).

[11] S.M. ULAM e I.C. OXTOBY - "On the existence of a measure in-
     variant under a transformation". Annals of Math. 40,
     560-566 (1939).

CENTRO INTERNAZIONALE MATEMATICO ESTIVO

(C,I,M,E,)

A. N. FELDZAMEN

THE ALEXANDRA IONESCU TULCEA PROOF OF MCMILLAN'S THEOREM

ROMA - Istituto Matematico dell'Università - 1960

THE ALEXANDRA IONESCU TULCEA PROOF OF MCMILLAN'S THEOREM

by   A. N. FELDZAMEN

The object of this lecture is to present a new, recent proof
of McMillan's Theorem, the first such proof for the case of convergence almost everywhere, and one which also includes convergence in
L^1 . It is due to Dr. Alexandra Ionescu Tulcea of Yale University.

The main idea of this proof is simple, elegant, and illuminating,
but I am almost certain it will not be published in this form.
Dr. Ionescu Tulcea later generalized the result and the methods, and
this generalization - rather than the present, somewhat simpler
and clearer, version - will be published shortly.

Professor Halmos has discussed the history of McMillan's theorem, and the work of Shannon, McMillan and Breiman, and we turn to
the details. A few preliminary definitions, and two theorems, are
required in this presentation.

We take fixed throughout a probability space $(X, \mathcal{F}, P)$, with
$P(X) = 1$. If $\mathcal{A}$ is a *finite* σ-field contained in $\mathcal{F}$, we write
$\pi(\mathcal{A})$ for the family of minimal sets, or atoms, of $\mathcal{A}$, and $C(\mathcal{A})$
for the cardinality of $\pi(\mathcal{A})$.

THEOREM 1. *If $\mathcal{A}$ is a finite σ-field contained in $\mathcal{F}$, and*
$\mathcal{C}_n \subseteq \mathcal{C}_{n+1} \subseteq \mathcal{F}$, *n = 0, 1,..., is an increasing sequence of arbitra-*
ry σ-fields, then

$$P(\{x \mid \sup_{0 \leq n < \infty} I(\mathcal{A}/\mathcal{C}_n)(x) > \lambda\}) \leq C(\mathcal{A}) \, e^{-\lambda}$$

PROOF : It is clearly sufficient to prove that for each $A \in \pi(\mathcal{A})$
we have

$$P(\{x \mid \sup_{0 \leq n < \infty} I(\mathcal{A}/\mathcal{C}_n)(x) > \lambda\} \cap A) \leq e^{-\lambda} \ .$$

A.N.Feldzamen

To prove this, let A be fixed and put

$$E(\lambda) = \{x \mid \sup_{o \leq n < \infty} I(\mathcal{Q}/\mathcal{C}_n)(x) > \lambda\} \cap A$$

$$F_o(\lambda) = \{x \mid P(A/\mathcal{C}_o)(x) < e^{-\lambda}\} \ ,$$

and

$$F_k(\lambda) = \{x \mid \inf_{o \leq j \leq k-1} P(A/\mathcal{C}_j)(x) \geq e^{-\lambda}, \ P(A/\mathcal{C}_k)(x) < e^{-\lambda}\},$$

for every $\lambda > 0$.

Then

(i) $\quad F_n(\lambda) \cap F_m(\lambda) = \phi \quad$ if $\ n \neq m \ ; \ n, \ m = 0, \ 1, \ldots$

(ii) $\quad F_n(\lambda) \in \mathcal{C}_n \ ; \ n = 0, \ 1, \ldots$. This is an immediate consequence of the definition of conditional probability.

(iii) We recall that, by definition,

$$\int_C P(A/\mathcal{C}) \ dP = P(A \cap C) \qquad \text{if} \quad C \in \mathcal{C} \ . \quad \text{Thus}$$

$$\int_{F_n(\lambda)} P(A/\mathcal{C}_n) \ dP = P(F_n(\lambda) \cap A) \leq e^{-\lambda} \ P(F_n(\lambda))$$

for each $n = 0, \ 1, \ldots$ and $\lambda > 0$.

(iv) I assert $E(\lambda) = \bigcup_{n=o}^{\infty} F_n(\lambda) \cap A$. For, if $x \in A$, we have

$$I(\mathcal{Q}/\mathcal{C}_n)(x) = - \log P(A/\mathcal{C}_n)(x) \ , \ n = 0, \ 1, \ldots \ . \quad \text{Thus}$$

$$x \in E(\lambda) \iff \sup_{o \leq n < \infty} (- \log P(A/\mathcal{C}_n)(x)) > \lambda$$

$$\iff - \log (\inf_{o \leq n < \infty} P(A/\mathcal{C}_n)(x)) > \lambda$$

(since the function $- \log$ is strictly decreasing in $(0, \ 1)$)

$$\iff \inf_{o \leq n < \infty} P(A/\mathcal{C}_n)(x) < e^{-\lambda}$$

As this last inequality is strict, there exists k such that $P(A/\mathcal{C}_k)(x) < e^{-\lambda}$. Then $x \in F_k(\lambda)$ for the smallest k with this

A.N.Feldzamen

property. Thus $E(\lambda) \subset \bigcup_{n=o}^{\infty} F_n(\lambda) \cap A$, and reverse inclusion follows in the same fashion from this chain of equivalences.

(v)   Finally

$$P(E(\lambda)) = \sum_{n=o}^{\infty} P(F_n(\lambda) \cap A) \leq \sum_{n=o}^{\infty} e^{-\lambda} P(F_n(\lambda)) =$$

$$= e^{-\lambda} P(\bigcup_{n=o}^{\infty} F_n(\lambda)) \leq e^{-\lambda}$$

and the theorem is proved.

THEOREM 2. *If* $T : X \to X$ *is a measurable (that is* $T^{-1} E \in \mathcal{F}$ *if* $E \in \mathcal{F}$ *), measure preserving (that is* $P(T^{-1} E) = P(E)$ *if* $E \in \mathcal{F}$ *) transformation,* $\mathcal{Q} \subseteq \mathcal{F}$ *is a finite* σ-*field,* $\mathcal{C}_n \subseteq \mathcal{C}_{n+1} \subseteq \mathcal{F}$ *,* $n = 0,1,\ldots$ *is an increasing sequence of arbitrary* σ-*fields, and*
$$\bar{h}^* = \sup_{o \leq n < \infty} I(\mathcal{Q}/\mathcal{C}_n), \text{ then}$$

(1) $h^* \in L^p(X, \mathcal{F}, P)$ *for every* p *,* $1 \leq p < \infty$

(2) $\exists$ *h measurable* $\mathcal{F}$ *such that* $h \in L^p(X, \mathcal{F}, P)$ *,* $1 \leq p < \infty$ *, h is measure preserving, and*

$$\frac{1}{n} \sum_{k=o}^{n-1} I(\mathcal{Q}/\mathcal{C}_n) T^{n-k-1}$$

converges to h almost everywhere and in the norm of each L^p *,* $1 \leq p < \infty$.

PROOF : (1)   let p be fixed, $1 \leq p < \infty$. Put

$$\left.\begin{array}{l} A_j = \{x \mid j < (h^*(x))^p \leq j + 1\} \\[2em] F_j = \{x \mid (h^*(x))^p > j \} \end{array}\right\} \quad j = 0, 1,\ldots$$

Then $F_{j+1} \cap A_j = \phi$, and $A_j = F_j - F_{j+1}$. Thus $P(A_j) = P(F_j) - P(F_{j+1})$ and

A.N.Feldzamen

$$\int |h^*(x)|^p \, Pdx \leq P(A_0) + 2P(A_1) + 3P(A_2) + \ldots$$

$$= P(F_0) - P(F_1) + 2P(F_1) - 2P(F_2) + 3P(F_2) \ldots$$

$$= \sum_{j=0}^{\infty} P(F_j)$$

$$= \sum_{j=0}^{\infty} P(\{x \mid \sup_{0 \leq n < \infty} I(\mathcal{A}/\mathcal{C}_n)(x) > j^{1/p}\})$$

$$\leq \zeta(\mathcal{A}) \sum_{j=0}^{\infty} e^{-j^{1/p}} \qquad \text{(using Theorem 1)}$$

$$< \infty$$

Thus $h^* \in L^p$.

(2) To prove the remainder, we must make use of deep results of ergodic theory. The first of these, an ergodic theorem for a sequence of functions, states the following :

If h_n, $n = 0, 1, \ldots$, is a sequence of measurable functions satisfying (a) $\sup_n |h_n| \in L^p$, ($1 \leq p < \infty$), and (b) $\lim_x h_n(x)$ exists almost everywhere, then there exists a measurable measure preserving function $h \in L^p$ such that $\frac{1}{n} \sum_{k=0}^{n-1} h_k(T^{n-k-1})$ converges to h almost everywhere and in the norm of L^p .

(See P.Maker, The ergodic theorem for a sequence of functions, Duke Math. J., 6, 1940; and also : L.Breiman, The individual ergodic theorem of information theory, Ann. Math. Stat., 28, 1957).

In this theorem let $h_n = I(\mathcal{A}/\mathcal{C}_n)$. Hypothesis (a) follows from part (1) of this proof, and hypothesis (b) from the fact that $P(A/\mathcal{C}_n)$ is a martingale (- see Sections 10 and 17 of Entropy in Ergodic Theory by Paul R. Halmos). Thus the proof is complete.

APPLICATION. In theorem 2 , let $\mathcal{C}_0 = \{\phi, X\}$, $\mathcal{C}_k = \bigvee_{j=1}^{k} T^{-j}\mathcal{A}$, $k = 1, 2, \ldots$ We recall that

A.N.Feldzamen

$$I\left(\bigvee_{j=o}^{k-1} T^{-j} \mathcal{a}\right) = I(\mathcal{a})\, T^{n-1} + \sum_{k=1}^{n-1} I\left(\mathcal{a}/ \bigvee_{j=1}^{k} T^{-j}\mathcal{a}\right) T^{n-1-k} =$$

$$= \sum_{k=o}^{n-1} I(\mathcal{a}/\mathcal{C}_k)\, T^{n-1-k}$$

ıs, by Theorem 2

$$\frac{1}{n}\, I\left(\bigvee_{j=o}^{k-1} T^{-j}\mathcal{a}\right) \to h$$

$n \to \infty$, both almost everywhere and in the norm of L^p , $1 \le p <$

ı the proof of McMillan's theorem is complete.